Beyond the Bulldozers: Can Green Initiatives Save Shrinking Cities

Alisha

Table of Contents

Chapter 1: Introduction

Background

Since the 1970s, cities in North America have been experiencing a new phase in urbanization dynamics. It is a form of reverse urbanization that has been termed urban shrinkage. The shrinking city phenomenon is believed to be a multidimensional process that affects the city and its suburbs (Pallagst et al. 2017). This phenomenon is characterized by population decline, economic recession, financial deficits, increasing unemployment rates, vacant housing units, and idled public facilities in the city center. Hence, shrinking cities can be defined as "urban areas that have experienced population loss, economic downturn, employment decline and social problems as symptoms of a structural crisis" (Martinez-Fernandez et al. 2012, 213). It is a worldwide phenomenon that has been linked to post-industrialization.

In the Northeastern and Midwestern United States, many cities have experienced this decline; however, some have developed policies to regenerate the dying central city. Many of these cities are actively combating the decline by launching creative initiatives aimed at addressing vacant and abandoned properties. For example, Philadelphia's Neighborhood Transformation Initiative (NTI) and Baltimore's Project 5000 each sought to demolish thousands of vacant structures in these cities' most distressed neighborhoods (Kromer 2002; McGovern 2006). Oftentimes, the goal is to establish green space, including new parks and community gardens. In other cases, cities have taken greening a step further by strengthening urban development policies to promote walkability for pedestrians, biking options for commuters, and neighborhood beautification. However, these redevelopment strategies have raised several concerns among some scholars of urban economics who see the demolition of properties for green spaces as a waste of resources (Hallet et al. 2019). An example of this is the criticism of the Imagine Flint plan of 2011.

Flint, a medium-sized city in Michigan, is one of many declining cities which has tried to redevelop by attracting private investors. This city prospered and grew with the expansion of General Motors at the beginning of the 20th century. However, overdependence on the giant manufacturer caused a huge problem when GM reduced its labor force from 80,000 to 8,000 people between 1978 to 2006. This led to out-migration and abandonment. In 2011, over 11,000 buildings were vacant and over 6,000 were left in a deteriorating state (Pallagst et al. 2019). In a bid to redevelop the city, government officials launched a master plan called "Imagine Flint" in 2011. This master plan introduced "green innovation areas" as a new type of land use. However, this did not sit well with residents who believed the government was downsizing their neighborhood for a cause that seemed uncertain. Another issue involves the maintenance of these green spaces, considering the shrinking economy of cities like Flint; it is critical to provide a maintenance plan otherwise these spaces can become eyesores. Critics of these strategies have called for the development of an innovative approach to address revitalization (Hallet et. al 2019)—hence the adoption of immigrant-friendly policies by municipal and local governments. However, these policies have also been challenged by conflicting Federal government policies (Shrider 2017).

Dayton is the sixth-largest city in Ohio and the county seat of Montgomery County. As of 2019, the total population of Dayton was 140,407 (US Census Bureau). Dayton belongs to a group of cities in the so-called Rust Belt that are characterized by their population decline and economic shrinkage over the years. Findings revealed that Dayton's population over the years has dwindled. Between 2000 and 2012, the city lost approximately 20,000 residents. However, the population of its foreign-born increased over this period by 1.8 percent (US Census Bureau). This increase could be attributed to their immigrant-friendly policies.

Like other Midwestern cities, Dayton experienced a rapid increase in population in the 19th century. The pull factor the city experienced during this time was a result of its manufacturing economy that attracted people from neighboring counties. In the early 20th century, Dayton gained even more economic prominence when it became one of the country's leading centers of innovation. Some of the major innovators included Edward Deeds and Charles Kettering who founded the Dayton Engineering Laboratories Company. Dayton's most famous inventors are the Wright brothers, Orville and Wilbur, who transformed the travel patterns of the world and made Ohio the "Birthplace of Aviation." Seventeen years after experiencing a flood that killed 361 people and caused damage estimated at $1.5 billion, the Great Depression took its toll on the city's economy causing it to lose 5000 workers in three years (8,500 in 1930 to 3,500 in 1933). By 1935 the total unpaid tax is equivalent to today's $150 million (Adam 2019). However, the city started recovering at the start of World War II when its strong manufacturing plants like Delco, Frigidaire, and others were awarded contracts worth hundreds of millions by the government to make bomb fuses, machine guns, rocket motors, and bomber landing gear assemblies. Despite this economic revival, the city continued to lose population. The decline of Dayton over the years has left the city a shadow of its former self. Not only has the city lost population, the city can now be characterized by its high number of abandoned properties.

Figure 1

Map of Study Area (in Orange). Cartography by Author.

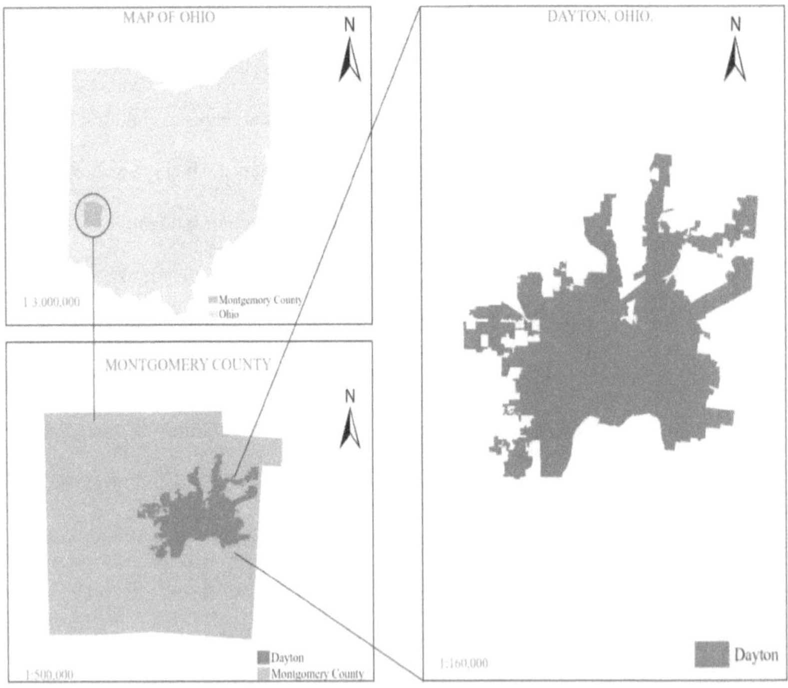

Purpose of the Study and Research Questions

Under the Trump administration, there was a dispute between the US federal government's hardline stance on immigration and states' and localities' immigration policies which are more welcoming. These debates over Trump's immigration policy overshadowed the fact that immigrants are just about 14% of the entire US population (Miranda and Majika 2019). Graauw suggested that immigration should not be the main concern for the Trump administration which banned migrants but instead should have used the country's attractiveness to migrants to its social and economic benefit (2017). Dayton exemplifies this more open strategy. Like many other Midwestern cities, it lost population in the past. But what makes Dayton different is that it recently started welcoming foreign-born people into its community—to the point that it is now regarded as one of the most immigrant-friendly in the country. The local government welcomed thousands of immigrants and accepted the change these immigrants bring through their "Welcome Dayton" initiative. Through this initiative, the municipal government has partnered with several organizations to help immigrants get jobs and provided easy integration.

This study investigates how migration is used to counter the shrinking of Dayton and how effective this strategy is. The study brings two bodies of research together: immigrant integration and urban revitalization, within the context of economic development and urban planning as part of a broader discussion about the role of immigration in reinventing a shrinking city. This research serves as a background for future studies related to immigration policies and disputes between local and federal governments regarding the potential for immigrants to benefit the economy. Lastly, it serves as an evaluation of Dayton's welcoming policy stance towards immigrants. The study will be shared with officials in the Human Resource Council who can then use it to evaluate the role of immigration in redeveloping the city while pinpointing areas where they can work further to ensure their goals of supporting immigrants and combating shrinkage are met. This study therefore provides answers to the following questions:

1. What is the role of immigration in planning for urban revitalization in Dayton, Ohio and how successful has it been?

2. More specifically, what impact have recent immigrants had on Dayton and what is the attitude of Dayton's non-immigrant residents towards these immigrants?

Chapter 2: Literature Review

Causes of Urban Shrinkage

Over the last decade, environmental researchers have delved into the study of shrinking cities and have come up with various definitions and causes of this phenomenon. Various criteria are used to explain the process and consequences of urban shrinkage including the life cycle of a city, population loss, and socio-economic factors like age, income, and unemployment amongst others (Meghan et al. 2019). Recent studies discovered that urban shrinkage since the mid-20th century has become a global challenge that has led to significant population loss in old industrial cities across North America, Europe, Asia, and Australia (Oswalt and Rieniets 2006). It is therefore easy to conclude that this phenomenon is quite universal. However, its determinants are less obvious and do not follow a universal standard. The complexity of policies related to deindustrialization, uneven investment, urban decentralization, suburbanization, and changing housing demographics are some of the causes of urban shrinkage (Megan et al. 2019). Rink et al. (2010) explained that the cause of urban decline in Detroit, USA could be compared to that of Donetsk, Ukraine, and Halle, Germany but further examination of these cities reveals remarkable differences. While Detroit's shrinkage is attributed to the story of "white flight," Donetsk's is a result of demographic change (increase in death rate), while that of Halle is a consequence of unemployment. Even though the cause of urban shrinkage differs from place to place, recent researchers like Haase et al. (2014) have noticed a universal symptom of urban shrinkage, which is population loss and economic decline. Different scholars like Beauregard (2006) and Haase et al. (2014) have theorized the possible causes of urban shrinkage in post-industrial cities which are discussed below.

This first school of thought believes that capitalists' investment in suburbs and disinvestment in cities is a catalyst in urban decline. These works of literature were grounded in the accumulation of capital and its spatial-temporal circulation, which focuses on the explanation of the dynamism of urbanization in capitalist society (Harvey 1982). Here, the main determinant of urban decline is not the people but rather the change in capital. Whilst some places are successful in attracting investment, others fail to do so and are plagued by abandonment, decreasing attractiveness, and, eventually, a decrease in residential population.

Harvey (1982) and Smith (1984) see the city as a "see-saw movement of investment, disinvestment, and reinvestment which makes uneven development a normal characteristic of capitalist urbanization" (Haase et al. 2014, 1522). Harvey further explained this phenomenon in his circuit of capital theory whereby focusing primarily on the process of production and the efficiency of the labor force to produce values and surplus value of which a proportion in turn sustains the reproduction of labor power (Harvey 1989). This theory places the importance of profit at the forefront of production since capitalists seek profit-making opportunities with little commitment to place. This theory then asserts that capitalism contributes to urban restructuring. To expatiate further on this, Beauregard (2006) in what he referred to as "parasitic urbanization," highlighted how suburbs grow at the expense of city centers. He explained how government policies have favored the expansion of suburbs while encouraging capital disinvestment in the city center and old suburbs. In the case of the Midwest cities, the majority of manufacturing firms relocated to the south or international locations where they can gain more profit (Vey 2008). Shepard (2016) identified how this theory has impacted the migration trend thereby contributing to the fast degeneration of cities. First, in the early stages of industrialization, rural-urban migration was experienced. This phase was characterized by the migration of low- and middle-class households into industrial cities. Secondly, the migration of both international and people of color into working-class neighborhoods led to tension among urban residents and accelerated the last phase in urban decline. The last phase is the urban-suburban migration of white middle-class who felt they had to vacate their abode because of the rapid growth of minorities in their neighborhood (Shepard 2016).

Another body of literature has identified how urban policies have contributed to displacement and urban decline. The post-World War II era saw the emergence of federal policies which aimed to remove urban blight from the inner city by implementing urban renewal projects under the Housing Acts of 1949 and 1954 (Hanon 2010; Shepard 2016). Although these projects aimed to revitalize the decaying part of the city, they ended up displacing people. An example of such a project is the Gratiot redevelopment project in Detroit. In an attempt to redevelop the city, the government partnered with private investors to build public housing which did not end up ameliorating the living conditions of residents but rather forced them to vacate their homes. Not only did some of these renewal projects displace residents, but they ended up destroying the 'inclusiveness' of the inner city which led to a gradual economic decline since capitalists preferred to invest in the suburbs because of how unattractive the inner city became after demolition projects (Shepard 2016). This was then followed by a sharp population loss since livable qualities and major sources of income disappeared from the inner city. Sugrue (1996) also described how housing policies like restrictive covenants and redlining were used to displace people of color. The restrictive covenants were used to ensure that blacks do not live in the same neighborhood as whites and it is usually achieved when "whites in a neighborhood would enter into private agreements with one another stipulating that none of their properties would be occupied, leased, sold or given to blacks" (Duneier 2016, 29). In Detroit, for example, real estate brokers and financial institutions did not allow blacks to own single-family houses in white neighborhoods; rather they were restricted to a deserted part of the city. Shepard (2016) argued that these discriminatory practices have given power to the advantaged while the minority groups are blocked out of local politics.

The construction and expansion of US interstate highways also contributed to urban shrinkage. Even though the expansion of interstate highways was aimed at easing the strenuous traveling condition between US cities, critics have argued that these expansion projects contributed to urban decline. There have been two major arguments around the expansion and construction of interstate highways. Scholars like Jones (2008) have argued that expansion projects have helped reduce commuting time. Jones (2008) supported his claim by tracing the significance of interstate highways back to World War I when the need to transport goods and munitions became a pressing issue. Kunstler on the other hand argued that the rise of automobiles has led to "the degradation of urban life caused by enticing the middle class to make their homes outside of town. It began an insidious process that ultimately cost America its cities" (Kunstler 1993, 90). An example is the Federal-Aid Highway Act of 1956. Popular at first, people began to see the negative impact of the 41,000-mile project when it inflicted damage on neighborhoods in its path, displacing people from their homes and dividing cities into different segments which according to Kunstler (1993) led to a gradual urban decline in the post-World War II era.

Despite being an urban decline catalyst according to some scholars, researchers like Emeric and Newman (2018) have also claimed that an efficient transportation network and the introduction of public mass transit could be used to reverse shrinkage as it encourages population stability. Emeric and Newman (2018) were able to identify a concrete relationship between public transport and population stability. In their words "The race for resources, access to and availability of reliable transportation have long been associated with human capital, an indicator of social stability" (Emeric et al. 2018, 342). This implies that people will not want to vacate their homes if they have access to an efficient public transportation system that helps them to commute easily. On the other hand, a poor transportation network might lead to a gradual out-migration and increase inequality. A typical case is the poor transport network of Dayton after World War II which forced the white population to migrate to the suburbs while leaving people of color in the declining city (Shepard 2016). In recent times, the introduction of rapid transit has had a positive impact on shrinking cities. An example is the substantial increase in both residential and non-residential buildings in the San Francisco Bay Area after the development of the rapid transit system (Cervero and Landis 1997). In the same light, Emeric and Newman (2018) were able to prove the significant relationship between the public transportation system and spatial distribution of neighborhood decline in Dayton. These scholars used a Geographic Information Science approach to show the influence of public transportation on urban development and It was discovered that neighborhoods that have easy access to transportation hubs did not experience a significant neighborhood decline compared to areas which lack access. Considering the various arguments, it is easy to conclude that public transportation can help reduce urban decline; however, before such projects are embarked upon, policymakers should first consult with residents on the best way to go about it.

Another theory attributes urban decline to demographic change which is solely based on the ratio of mortality to birth rate. Haase et al. (2014) argued that a continuous decrease in birth rate without decreasing mortality rate in some European countries would result in a long-term population decline. Connected to this, the changing patterns in population concentration that Long and Nucci (1997) investigated have more recently been shown to be connected not only to regional migration patterns but also a heterogeneous geography of births and deaths (Rogerson and Plane 2013). Thus, there is a consensus that urban shrinkage is synonymous with population loss. Most of these researchers have identified several factors responsible for this population loss such as unemployment (Harvey 1982). However, going through the literature very few of them have been able to recognize the trickle-down effect where housing stock found in the core city, which is more affordable to low-income earners, deteriorates faster since low rents induce landlords to neglect maintenance and upkeep. This effect could also contribute to urban shrinkage since residents would get to a point where they would have to move out due to neighborhood physical decay.

Another major cause of urban decline and population loss is the inevitable life cycle of cities. This theory has been proven to be a major determinant in urban renewal (Metzghar 2000). It has its root in the context of urban planning where Scholars like Babcock (1932) asserted that the decline of a neighborhood or city is inevitable. This theory was explained in five stages (see Figure 2) and has evolved over the years (Metzghar 2000). First is the development stage which is characterized by the development of new structures like the single-family residential development. This stage is then followed by the growth phase where everything seems to be normal, and the city is attracting immigrants. At the transition stage, things begin to change gradually due to the aging infrastructures especially when they are not properly maintained. Also, the population increase stretches infrastructures beyond their limit. This is then followed by the decline stage otherwise known as shrinkage where investments and rent values depreciate, and people then migrate out of the city, abandoning their homes. The renewal/redevelopment phase is then introduced to replace deteriorated buildings.

In the same light, Berry (1977) and Van de berg (1982) argued that the major reason for population loss in cities is natural. They describe what we now refer to as shrinkage as an aftermath of counter-urbanization. They both focused on the population loss in city centers, which was attributed to the pressure on amenities within the center which they believed led to an out-migration to the city suburbs. Both scholars developed models to further solidify their findings and it can be concluded that they both believed that shrinking is an inevitable experience in a growing city. These researchers have raised salient points to explain one major cause of urban shrinkage in our cities that other researchers did not pay attention to. However, they did not provide measures to counter this inevitable factor.

Figure 2

Life-cycle theory of a city (Metzghar 2000).

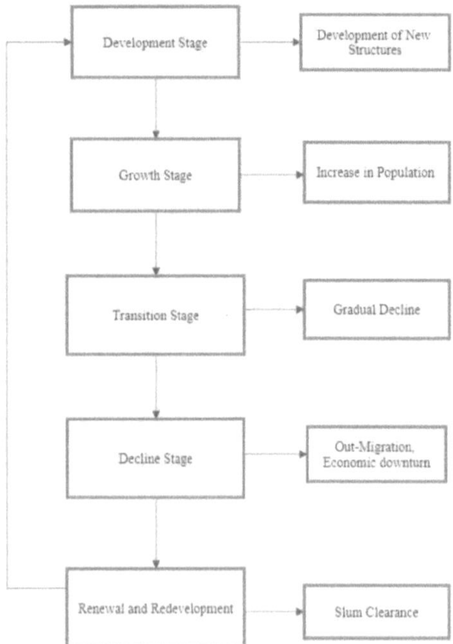

Urban Regeneration and Sustainability

In recent times, the term sustainability has evolved beyond the definition of the

United Nations – "meeting the needs of the present without compromising the ability of

future generations to meet their own needs" (United Nations 1987). Several researchers

have modified this definition to place priorities on our home— "putting in place the age-

old desire to live richly, equitably, and peaceably with each other and with our natural

surroundings so that our 'home' may offer itself to the imaginations and tables of

generations to come" (Bernard, quoted in Buckley 2014, 374). Rapid urbanization has

called for urgent implementation of sustainability measures. However, sustainability has

been used for selfish gains among industrialists. As suggested by Buckley (2014), the

term sustainability no longer holds value; instead, it has been used by capitalists to

generate profits. The generic usage of the term sustainability in a capitalist environment

has forced environmentalists/planners to adopt the new term 'thrivability' which they

believe has more depth than sustainability. According to Buckley, "it offers a more

optimistic approach, one that pushes us toward a future we desire as opposed to avoiding

a future that frightens us" (2014, 378). Considering this, several environmental advocates

like Cohen (2018) and Thompson (2018) conclude that doing less harm to the

environment is no longer enough, but instead, we should move towards the regeneration

of sustainability. Thompson (2018) defined this regeneration as restoration of what we

have lost and making sure that we build a sustainable economy that will flourish.

Several environmentalists have raised concerns about whether we can achieve

urban sustainability and have suggested different approaches which may lead us on the right path. Urban regeneration is one of the key themes that has been suggested by researchers in various fields. The framework of urban regeneration cuts across "revalorizing urban space" (Shaw and Porter 2009, 2 cited in Shepard 2016). There are many definitions of urban regeneration in the planning literature; however, it is difficult to find a definition that touches all the themes in urban regeneration. One definition explained that urban regeneration involves a "comprehensive and integrated vision and action which seeks to resolve urban problems and bring about a lasting improvement in the economic, physical, social, and environmental condition of an area that has been subject to change or offers opportunities for improvement" (Roberts 2000, 17). Roberts went further to explain how urban regeneration could be achieved by the adoption of policies that aim to provide a long-lasting solution to not just environmental problems but also economic, physical, and social issues. However, sustainable urban regeneration is not limited to this. It also aims to preserve natural resources for long-term development while minimizing the disruptive effects of urbanization on the natural environment (Evans 2012; Wheeler and Beatley 2008). For Urban regeneration to be a success, Affernuzamman et al. (2018) stressed that stakeholders at each level of authority must share the same goal. Given the different perspectives of scholars and the scope of urban regeneration, Shepard (2016) identified six approaches that should be included in urban regeneration to ensure a sustainable revitalization. First is the property-led approach where it is believed that "properties should have multiplier effects in the local economy" (Shepard 2016, 17). The second strategy looks at the business-driven approach which looks at how abandoned markets/industries could be used to revitalize the inner city. Next is the cultural industries approach which focuses on how the creative and media industries could be developed to serve as a revitalization tool. "Cultural regeneration involves the preservation of industrial heritage buildings together with measures of economic revitalization through their transformation into active cultural spaces that pursue to display the patrimony assets of industrial technique and identity" (Merciu et al. 2017, 415). The fourth strategy places much emphasis on the importance of sustainability in new design perspectives of the city and its urban form. This is followed by the well-being perspective of residents. Here it is believed that a well-designed space which includes gardens and parks would promote good physical and mental health of residents and prevent possible out-migration to suburbs. Lastly, Shepard (2016) identified the community-based strategy which simply put is the bottom-up approach in decision making where residents are represented in decision making. This allows policymakers to get a view of residents' needs and how these needs can be met.

Researchers have come to the understanding that urban regeneration not only reduces the harsh impacts of urban development on the natural environment; they also concluded that it comprises a variety of policy frameworks that could be adopted to redevelop the socio-economic and spatial organization of dying cities. In agreement with this, Tallon (2010) inserts sustainability into the frame of urban regeneration and concludes that sustainable regeneration does not only aim to improve urban quality of life but also prevents displacement of local people, supports local identities, provides social and regional balance, and encourages participation and partnership of government and non-governmental stakeholders. Recent examples of sustainable urban regeneration practices have an explicit climate change dimension. In some countries, recent regeneration projects aimed to create resource-efficient and climate-friendly urban quarters where global warming potential and ecological footprints of urban development have been lowered (Balaban and Puppim de Oliveira 2014). However, it should be noted that sustainable urban regeneration can only succeed when social, environmental, and economic dimensions are treated equally. To do this, urban regeneration projects must be developed as part of a comprehensive urban development program and implemented within a public participation process (Shepard 2016).

Urban Shrinkage and Smart Decline in US Cities

Shrinking cities can be defined as "urban areas that have experienced population loss, economic downturn, employment decline and social problems as symptoms of a structural crisis" (Fernandez et al. 2012, 213). In the same light, Wiechmann (2007) defined shrinking cities as "a densely populated urban area with a minimum population of ten thousand residents that have faced population losses in large parts for more than two years and is undergoing an economic transformation with some symptoms of a structural crisis." Some of the features of a declining city include abandoned or unoccupied buildings, high crime rates, empty overgrown lots, and aging and indigenous populations (Pallagst and Shwarz 2009). Urban shrinkage has affected the living conditions of residents in different ways. It is a fact that population loss and urban shrinkage are inseparable. Adams (2019) gave an example of how population loss leads to revenue shortage and poor quality of life. Adams explained how Dayton has lost an equivalent of 150 million dollars in today's economy from tax revenue. This shortage of funds can strain a city's ability to maintain existing infrastructure. To address this, the government may then be put in the position where they must impose higher taxes on the remaining residents (Heim et al. 2019). This approach can then force remaining residents to migrate when taxes become unbearable. To address the negative impact related to population loss several urban planners and environmentalists are agitating for the adoption of new planning policies. These policies aim to realign public infrastructure and redistribute public resources in such a way that it meets the needs of the current status of the city (Rybczynski and Linneman 1999). These new policies are what academics like Popper (2002) have described as smart decline.

Smart decline is defined as "planning for less people, fewer buildings, fewer land uses." (Popper 2002, 23). Popper further described it as "basically leaving behind assumptions of growth and finding alternatives to cities with declining population and jobs" (Park 2018, 55). This definition stems from the fact that a lot of declining cities in the late 1990s did not come to the reality that their shrinkage would be difficult to reverse. This could be better explained using the example of Midwest cities which to a large extent relied heavily on automobile industries and unfortunately did not pay adequate attention even when these factories started losing their relevance in the automobile industry. Instead, some of these cities' officials were in denial that they were experiencing shrinkage and believed they could sustain their population. However, the surplus abandoned properties discouraged reinvestment which further led to market dysfunction and frustrating the revitalization of both public and nonprofit organizations (Rybczynski and Linneman 1999 in Schilling and Logan 2000).

Urban scholars like Pallagst and Popper have identified some smart shrinking strategies that could help redevelop these cities. First is the right sizing and greening approach. "Right-sizing refers to stabilizing dysfunctional markets and distressed neighborhoods by more closely aligning a city's-built environment with the needs of existing and foreseeable future populations by adjusting the amount of land available for development" (Schilling and Logan 2000, 453). There are different approaches to the right-sizing of a shrinking city. One is de-annexation, which is aimed at restructuring the boundaries of the municipal area in a way that urban services are reduced to the needs of the remaining population. Another right-sizing strategy is the removal of unneeded public infrastructure from the city. This helps municipal governments to reduce maintenance costs, focusing on providing services to areas where it is needed most. Also, to reduce cost, Schilling and Logan (2000) suggested that municipal governments shift service responsibility to private investors. Despite its efficiency in revitalization, right-sizing has raised issues of social equity because vacant neighborhoods are often dominated by people of color and low-income earners. To address this issue, planners must ensure that the needs of the remaining residents are equally balanced, and they should ensure it's a bottom-up approach where residents are actively involved in development plans and relocation alternatives. The redevelopment success of Philadelphia's Neighborhood Transformation Initiative and Youngstown's Comprehensive Plan could be measured to explain the importance of residents' involvement in right-sizing policymaking (Schilling and Logan 2008). In Philadelphia, the municipal government did not pre-inform residents on the benefit of demolition. Neither did they effectively address their fears of relocation.

As a result of this neglect by the government and the fear of being kicked out of their homes just like in the days of urban renewal, residents revolted against the NTI plan after two years of its operation. Even though the government readjusted by providing full benefits to relocating residents under the Federal Uniform Relocation Act, the NTI still ended up not meeting its goal of reclaiming over 14,000 abandoned properties.

The Youngstown Comprehensive Plan, however, has been a success because of how the public was carried along in every formation stage (Faga 2006). At first, community leaders in Youngstown believed they could recover over half of their lost population since 1950 and the city would return to the vibrant life it once had. However, in 2005, the city came to accept its fate and called for a "better, smaller Youngstown focusing on improving the quality of life for existing residents rather than attempting to grow the city" (Hollander and Nemeth 2011, 6). In the formation process, the planning committee ensured that the public was involved by hiring media outlets to run a citywide campaign that explained the benefits of the plan. This campaign recorded success as residents were able to voice their opinion on websites created by the media outlet and over 1,300 residents attended the first presentation of the plan in 2005 (Faga 2006). Since the adoption of this plan, it has gained huge popularity and positive responses from both academics and media outlets. The *New York Times Magazine* went so far as recognizing the plan as one of the 74 best ideas in America (Rhodes and Russo 2013). The master plan was applauded for its smart decline approach which is visible in the Buffalo Commons proposal where they looked at revitalization by preserving a large portion of the Great Plains for ecological and economical restoration (Rhodes and Russo 2013).

City officials believed this approach was the best to combat population and economic decline. In the same light, Reimer et al. (2015) argued that greening, green infrastructure, and green spaces have become a wide applied urban development tool not only for shrinking cities. Thus, greening can be viewed as a strategy on all planning levels from state to city. Pallagst et al. (2019) concluded that declining cities that have adopted these strategies have created new trends for sustainability and the quality of life for their citizens.

Shilling et al. (2008) and Adelaja (2010) both suggested that shrinking cities containing brownfield sites could revitalize these spaces for a variety of uses, be it parks, urban gardens, new habitats, flood-mitigation, and/or agricultural areas. The definition of brownfield sites has evolved over the years. It is no longer seen as just abandoned/dilapidated industrial buildings. It has also expanded to include "not only contaminated lands but all previously developed land available for reuse with or without intervention" (Maliene et al. 2010, 2). The concept of reusing brownfield sites has been widely accepted in many countries. For example, the federal government in Germany has promoted the conversion of abandoned industrial sites in the inner city into mixed uses, which are environmentally friendly, and cost-effective. Environmentalists like Greenberg et Al. (2001) have referred to brownfield redevelopment as a smart growth option which aims to attract population and investors back into the city. However, to successfully achieve this, they suggest the government should:

(1) Make the redevelopment of neighborhoods with brownfields an unambiguous bipartisan priority.
(2) Locate or expand government facilities in these places.

(3) Provide incentives to private investors who locate in desired neighborhoods, including funds to remediate sites and demolish buildings, tax reductions, and liability protection.

(4) Make it financially painful for businesses and speculators to retain brownfield properties in an unused or underutilized mothballed state.

(5) Provide funds to upgrade and/or add infrastructure, police, fire, sanitation, social and public health, and other key services to improve the quality of neighborhoods with brownfields.

(6) Reach out to individuals who are likely to be attracted to these locations with promotional messages, maps of sites, and other information that touts the advantages of these locations, including listing government-supported incentives. (Greenberg et Al. 2001, 130).

A success story of this brownfield redevelopment strategy can be found in St. Paul, Minnesota where the Hmong community successfully converted an abandoned 3.3-acre industrial waste dump into a funeral home (EPA 2010). In response to the criticism of this strategy that health hazards might stem from reusing old factories, the Environmental Protection Agency (EPA) has made grants available for the clean-up process. An example of such a grant is the $400,000 clean-up grant that was awarded to the Vang Pao foundation in St. Paul after discovering high levels of hazardous chemicals in the waste dump (EPA 2010). In the same light, the state of Ohio has taken a huge step towards the redevelopment of brownfield sites and has provided loans through the 'Ohio Brownfield Fund' which provides loans between $500,000 and $5,000,000 for environmental assessment and environmental clean-up, respectively. Despite the numerous works of literature on the success stories of brownfield reuse, very few have pointed out the possible impact of this reuse on ecology and public health (Greenberg et al. 2001). They went on to identify both the positive and negative impact brownfield redevelopment has on air, water, disaster prevention, and public health. They concluded that, with the help of environmental impact assessment and public participation in redevelopment strategies, most of the negative impact could be ameliorated.

Immigrant-Driven Revitalization Policies

Following the deindustrialization in Midwest cities, leading to loss of population, loss of manufacturing jobs, and economic downturn (see table 1), these cities have digressed from the traditional revitalization strategies by becoming intentionally immigrant friendly. In a bid to revitalize, these cities have tried different traditional strategies aimed at attracting investors and the population. Some of these strategies include the provision of incentives like tax rebates to potential investors, and the construction of new infrastructure for the company's specific benefit. (Shrider 2017). To generate more revenue, some have also invested in the redevelopment of their downtown into an entertainment district. Although some of the revitalizations were successful, many of these cities are still experiencing social and economic problems. It is in this context, that a lot of these cities have decided to welcome immigrants and refugee populations in a move to upgrade their economies. Although these cities have claimed not to have been motivated by the advantages that come with the presence of immigrants, Shrider (2017) argued otherwise that these policies were indeed aimed to revitalize their cities and their economies.

Table 1

Employment Status of some Midwest Cities in 1960 and 2010 (Shrider 2017, 10-11)

City	Population 1960	Population 2010	Population Change (%) 1960-2010	Percent Employed in Manufacturing 1960	Percent Employed in Manufacturing 2010	Percent Unemployed 2010
Akron, OH	290,351	199,110	-31.42	43.99	15.67	12.9
Cincinnati, OH	502,550	296,943	-40.91	29.4	11.24	10.7
Cleveland, OH	876,050	396,815	-54.7	40.81	13.58	17.8
Dayton, OH	262,332	141,759	-45.96	36.4	12.64	16.1
Gary, IN	178,320	98,294	-54.97	50.49	13.44	19.1
South Bend, IN	132,445	101,168	-23.62	39.43	16.88	13.5
Youngstown, OH	166,689	66,982	-59.82	40.96	13.42	18.4

Shepard (2016) mentioned that this new revitalization approach is not restricted to the Midwest alone, but rather it is spread across the United States although under different names. These immigrant-friendly initiatives rely on the never-ending influx of immigrants and refugees into the United States. This phenomenon has given cities the opportunity of including immigrants in their revitalization plans (Shepard 2016), hence the term 'immigrant-friendly' city was generated. Immigrant-friendly has been defined as a community "that has a real honest infrastructure that can provide the kind of empowering services that people need. Not one that calls itself a welcoming city and doesn't have or chooses not to put resources behind the initiative." (Shepard 2016, 121). Shepard went on to say that before a city can be termed welcoming, policymakers must ensure that the immigrant community is involved in decision-making. In other words, immigrants should be treated like native residents. Successful immigrant-friendly cities have reaped the benefits associated with this initiative and scholars like Singer and Wilson have explained some of the benefits. First, is the economic growth of these cities through augmentation of the labor force and their contribution to the economy as final consumers of goods and services (Vigdor et al. 2013). Herman and Smith (2009) also identified that immigrants are 30% more likely to be entrepreneurs than native-born and often invest in desolate areas where it is cheap to run a business. Singer and Wilson (2006) also identified that the influx of immigrants and refugees has turned around the population loss in declining cities. Thirdly, immigrants have helped remodel the face of declining neighborhoods through the renovation of abandoned buildings. Researchers have proved that immigrants are often attracted to small cities with affordable housing and fortunately this is one of the characteristics of shrinking cities. Once immigrants get into these cities, they acquire and renovate these properties. Despite the consensus that being immigrant-friendly is an effective strategy for declining cities, implementation of these policies is often challenged by different factors, some of which are explained in this section.

Portes and Rubaut argued that the migration pattern of a country is influenced by its context of reception which is based on three factors—government policies, labor market, and existing immigrant community (2006). Municipal government initiatives are difficult given the fact that the government only has direct control over one aspect of the three factors. Shrider (2017) however mentioned that municipal governments have tried to improve their local immigration policies by moving from either passive or exclusion acceptance to active encouragement. To further expatiate on what these different types of acceptance mean, Fleury and Longazel (2010) explained that exclusion acceptance refers to government policies that are meant to discourage immigrants from settling in their cities. Arizona's SB 1070 is a perfect example of where this unfriendly policy was adopted. In this state, the police department was saddled with the responsibility of checking immigrants' documents (Archibold 2010; Shrider 2017). Passive acceptance is a neutral ground between the three types of immigrant policies. Here, the municipal government does not show interest in either accepting or rejecting new immigrants. They maintain their current population and adhere to the federal government immigration policies instead of forming their own. This implies that the flow of immigrants to such cities will solely depend on the other two contexts of reception—labor market and immigrant community. Lastly, active encouragement refers to the policies enacted by local government to attract immigrants into their cities. To achieve their goal, local governments often develop their immigration policies in which they include promotional incentives like healthcare for immigrants. This type of policy has been adopted by Midwest cities like Dayton, Ohio. In concluding this section, it is obvious that shrinking cities would most likely find it hard to attract immigrants because they only have control over one (immigrant community) of the three contexts of reception. Their loss of manufacturing jobs has deprived them of labor force factors. Likewise, the unfriendly immigrant policies of the federal government have frustrated these immigrant friendly cities from reaching their goal.

Another challenge in implementing immigrant-friendly policies is the tussle between federal and local immigration policies. Tracing the immigration history of the United States revealed that the federal government has the sole authority to control immigration policies giving no room to state and local governments to create their own immigration policies. (Hummel 2015). Local governments were only given the right to create policies that speak to how they want to integrate immigrants into their cities. This political structure has frustrated immigrant-friendly cities as they must comply with federal policies which in recent times haven't been friendly to newcomers. Trump's immigration policies are an example of these federal-level, unfriendly policies which did not allow local immigration policies to flourish.

Lastly, criticism from residents has also been an impediment to the goal of immigrant-friendly cities. Hummel (2015) explained how immigrants are often linked by non-immigrants to crime in the city. He used the opportunity structure theory to prove that immigrants only commit crimes in cities when they are not given opportunities to create wealth. However, researchers like Adelman and Williams (2017) have refuted any connection between immigrants and increased violent crime with their findings that crimes like larceny, murder, and robbery have reduced in number in the largest 200 metropolitan cities between 1970 and 2010 (Adelman et. Al. 2017: Shrider 2017). Another challenge, however, is that some residents argue that the influx of immigrants in their cities has led to a shortage of employment available to them. Some residents also resisted these immigrant-friendly policies because they believe the influx of immigrants has saturated their labor market leading to the reduction of their wages (Hummel 2015). Nevertheless, some cities like Dayton continue to pursue an immigrant-driven revitalization approach.

Dayton's Revitalization Plan

To revitalize the city, Dayton officials have adopted several approaches/policies. In 2013 the *New York Times* published Julia Preston's article titled "Ailing cities across the Midwest extend a welcoming hand to immigrants." The post revealed how declining cities have been welcoming immigrants and much attention was given to Dayton, which has welcomed hundreds of Turkish migrants. Despite its struggles with deindustrialization, city officials aspire to make Dayton the "Silicon Valley" of Ohio. To achieve this, they have put in place strategies that aim to redevelop the downtown, attract businesses, and increase housing stock (they aim to build about 4000 affordable housing units). Expanding a conference center and surrounding hotels and creating a minor league baseball stadium and a concert hall were also included in their redevelopment plan. This development plan has yielded positive outcomes and attracted immigrants. The Turkish community in Dayton has grown from one hundred to over two thousand in recent years and they have been well integrated into the system (Navera 2019). Not only has this community populated Dayton, but it has also contributed to the social and economic development of the city. Some of the pull factors include the low cost of houses in North Dayton, low cost of living, educational facilities around the city, and connection with relatives or friends. The Ahiska Turkish community in Dayton always encourages other Turkish immigrants to move to the city (Shepard 2016). City officials also adopt policies to welcome more immigrants along with its other initiatives.

The planning strategies outlined in Dayton's comprehensive plan also represent "maintenance" planning strategies, which focus on rehabilitating and redeveloping existing housing and infrastructure, enforcing codes related to building maintenance, and utilizing infill development and historic preservation. The plan also includes some key "smart decline" planning strategies—for example, establishing more regional partnerships and collaborations as a cost-savings measure, downzoning as to have current zoning match existing physical characteristics of neighborhoods, and engaging in land banking and community land trusts. Their companion plan, Reinventing Dayton and the Miami Valley through Vacant Property Revitalization and Reclamation, details specific planning strategies aimed at addressing vacant properties, including demolition and condemnation, rehabilitating properties, and housing stabilization programs aimed at preventing maintenance and code issues, abandonment, and vacancy (Dayton, OH 2005).

Chapter 3: Methodology

To understand the impact of becoming an immigrant-friendly city, this research adopted the ethnographic methodology approach where I used qualitative research tools including semi-structured interviews (primary source) and analysis of relevant documents (secondary source). To answer my research questions, I decided to adopt the interview methodology because it involves oral exchange which will help fill a gap in knowledge where other traditional techniques might fail. Dunn (2010) stressed further that "interview is an excellent method of gaining access to information about events, opinions and experiences" (2010, 150). The semi-structured interview was selected as the most appropriate interview method for this research because of its flexibility. A semi-structured interview requires an interview guide that would contain relevant questions that are related to my research questions. I chose this type of interview because despite preparing carefully worded questions in the guide, this approach allows the interviewee to set the direction of the interview.

Semi-structured interviews were used to obtain information related to immigrants' experience and knowledge of the city, how they have been treated by native residents, and how they have benefited from the immigrant-friendly environment. Policymakers were also able to divulge the impact of immigrants on the city, their involvement in policy formation, how their immigrant-friendly initiatives work, and their achievement. With the help of these semi-structured interviews, I was able to learn about personalized immigrant experiences.

Field notes were also instrumental during this research as they helped keep track of cross-sectional ideas which emerged from my interviews and document analysis. To write comprehensive fieldnotes, I started by jotting down my experiences during the interview before developing them into fully developed fieldnotes. According to Emerson et. al (2010), fieldnotes help researchers to remember distinct discourse that came up in interviews. Dunn (2010) also mentioned that fieldnotes help researchers to develop their personal and analytical reflections which are useful in identifying patterns in interviews. My reactions and personal reflections revolved around the struggles of immigrants, their impact, their communal relationship, and to what degree the municipal government has supported them.

Recruitment and Sampling Process

This research made use of two classes of participants: municipal policymakers and immigrants in Dayton. The policymakers were selected based on their involvement in the formation of immigrant-friendly policies and partnerships with immigrant communities. The immigrants on the other hand were selected based on the number of years they have lived in Dayton. I believe immigrants who have lived there for a longer period would have more experience hence I selected immigrants who have lived over three years in Dayton. The pandemic situation in the country at the time of this research made it quite difficult to access most of the offices I had initially identified as potential data sources. I therefore relied on the Human Resource Council's website to contact those whose job title fits my research interest and luckily, the HRC website had a dedicated page to Welcome Dayton Staff out of which I emailed some staff.

After reaching out to my first set of participants via phone call or email depending on their convenience, I then relied on a snowball sampling method. This method is useful in reaching out to participants who are not easily accessible, and it also allows them to communicate better with the researcher based on their relationship with the first participant (Naderifar et al. 2017). Once I completed an interview with each participant, I asked for suggestions of people I could reach out to for research purposes. Thereafter I sent an email to these new contacts using the updated version of the recruitment document submitted during my Institutional Review Board (IRB) application (see Appendix A). I ensured this document followed the guide of Dunn (2010) which says a recruitment document must contain a short introduction of the researcher, how he/she got their contact, the purpose of reaching out to them, why they were selected for this research, the significance of the research, and lastly, duration of interview and contact details.

Before the study, I had envisioned that the snowball sampling method would help me connect to immigrants from diverse ethnic groups and policymakers from all the sub-committees involved in Welcome Dayton. Sadly, this did not happen. Nevertheless, with this snowball technique, I was able to reach out to thirteen (13) potential participants out of which seven (7) responded and agreed to be interviewed. Out of the seven participants, four were decision-makers who were actively involved in the formation of the Welcome Dayton plan while the remaining three were Turkish immigrants. Table 2 shows the participants' information.

Table 2

Background Details of Participants and interview procedure

Participants	Pseudonym	Gender	Occupation	Interview Procedure
1	Katherine	Female	Human Resource Staff member who was actively involved during the formation of Welcome Dayton policy.	Phone Interview
2	Ruiz	Male	Government official, he was actively involved in the formation of Welcome Dayton	Phone Interview
3	Linus	Female	Government official who is also a member of Welcome Dayton Committee	Phone Interview
4	Tetula	Male	Government official who was involved in Welcome Dayton formation and aware of other revitalization plans in addition to being immigrant-friendly	The interview was over zoom video call
5	Dovian	Male	Turkish Business owner who has lived in Dayton for over 10years.	The interview was done over zoom video call

Table 3 continued

Background Details of Participants and interview procedure

Participants	Pseudonym	Gender	Occupation	Interview Procedure
6	Lamidi	Male	An immigrant who is a member of the Welcome Dayton Committee and has lived in Dayton for over 20 years	Phone Interview
7	Yun	Male	Turkish Immigrant who has lived in Dayton for 2 years. He is a student who already has a menial job.	Phone Interview

Interview Procedure and Transcription

This research made use of two different interview protocols—one for the immigrants and a second one for policymakers. Both protocols adopted the semi-structured interview format which is a middle point between the structured and unstructured interview. This format was chosen because of its flexibility which allows me to further probe my participants when I sense that I could get more information based on their experiences. Also, this type of structure gave me the liberty to modify my questions in a way that allows me to get significant information from my participants. Each interview protocol for this research contained about twelve open-ended questions which have multiple sub-questions. I ensured that each interview protocol started with easy questions which were basically to familiarize myself with the participant and also to reduce anxiety as suggested by Dunn (2010). I also allowed participants to express themselves by rewording some of my questions and modifying my protocol while making sure the conversations were in context with the research goal.

Due to the pandemic, all my interviews were conducted over the telephone and zoom video call (see table 2). Before the scheduled interview date, I emailed a copy of my IRB consent form to each respondent. This form explains what the research is about, how recordings would be kept safe, and emphasized that their identity would remain undisclosed. It is important that interviews are transcribed accurately in this research. Hence, each interview was recorded after obtaining consent from the participant. Al-Yateem (2012) pointed out that audio recording during interviews could make respondents censor information as they become self-conscious and avoid saying statements that could implicate them. To cushion this effect, I worked to gain their trust by having conversations that are outside my research before starting my interview. We often talked about my stay in the United States and how I have been coping as an international student. Talking about this made immigrant respondents more comfortable. I also assured them that I would keep their information confidential.

Transcription is an essential part of my analysis and this was done almost immediately after each interview. Audio recordings were transcribed verbatim using the OTranscribe online software which allowed me listen to the interview while simultaneously typing the transcript. Not only is this software free it also has functions that allow me to reduce the speed of the recording and also rewind if need be. Transcribing by hand was a long and difficult process. However, doing this myself gave me the privilege of including non-verbal sentiments of my interviewees which I noticed during my in-person interviews. Dunn (2010) also mentioned that transcribing by hand allows the researcher to familiarize himself with the data which serves as a form of preliminary analysis. Throughout the transcription process, I kept an in-process memo which helped me retain my line of thought and kept my ideas in check. Birk et al. explained the benefit of memoing when they said, "memos can help to clarify thinking on a research topic, provide a mechanism for the articulation of assumptions and subjective perspectives about the area of research, and facilitate the development of the study design" (2008, 69). In this research, memos were kept in a notepad and were further developed as I got into the depth of the research.

To triangulate my data, I combined my research method (semi-structured interview) with document analysis. Bowen (2009) defined document analysis as "a systematic procedure for reviewing or evaluating documents-both printed and electronic material" (Bowen 2009, 27). This research method was used because the information contained in documents can suggest some questions that need to be asked and situations that need to be observed as part of the research. Also, because it can be analyzed to verify findings or corroborate evidence from other sources. (Bowen 2009). However, because these documents are often produced for other purposes outside research, the information that can be obtained from this method is often limited. Therefore, I only used this method for corroboration purposes and a base upon which I built more questions in my interviews. The majority of the documents used for this research were the meeting minutes from the Welcome Dayton committee. These documents were analyzed because of their significance to my research interests. It also allowed me to read the different opinions of the Welcome Dayton committee members.

Data Analysis

After transcribing all my interviews, I then adopted the qualitative coding techniques. Chamz defined qualitative coding as the process of "naming segments of data with a label that simultaneously categorizes, summarizes and accounts for each piece of data" (Chamz 2006, 43). One of the purposes of coding is to create a well-organized structure that helps us make meaning of our qualitative data (Cope 2010). The different types of coding techniques include descriptive coding, attributive coding, and in vivo coding amongst others. However, the purpose of this study and the type of data collected made analytical coding the most appropriate. Analytic codes "reflect a theme the researcher is interested in or one that has already become important in the project" (Cope 2010, 283). For this study, I coded in two stages—initial (open) coding which was then followed by focused coding.

The open codes that I used reflected themes or patterns which were obvious on the surface or stated directly by participants during interviews. Most of my open coding was done in Microsoft Word where I identified what information was being passed by participants and how it relates to my research goal. I left comments on the document which explained what I could make of the conversation. Emerson et al. (2010) advised that researchers should come up with as many codes as possible even if it does not fit into the theoretical focus of the study initially. To generate enough codes, I used the line-by-line coding technique. Even though it was time-consuming and at some point, seems arbitrary as some sentences were not completed, this technique ensured that all conversations were coded. These generated codes were then developed into a broader thematic idea which is known as the focused codes. Focused coding involved "using the most significant and/or frequent earlier codes to sift through large amounts of data" (Charmz 2006, 57). In this stage, I had to read my transcripts and documents again and then decided what sections of the text are related to individual-focused codes. For this process, I employed the use of a computer-assisted Qualitative Data Analysis Software (CAQDAS) known as Nvivo.

Chapter 4: Making a Declining City Immigrant-Friendly

The major contributor to the population declines in Dayton is the disinvestment and relocation of economic activities to other areas. Dayton, which was once a hub of industry in the Midwest, began to experience decline when huge companies like General Motors, Mead Paper, and others were forced to shut down and relocate their plants due to the gradual economic shift to service industries. These employers left behind thousands of unemployed residents who were then forced to migrate to the suburbs in search of jobs. This suburban population growth reduced Dayton's population from over 260,000 in the 1960s to 140,000 in 2010. In a bid to survive and revive the city, government officials have come together to develop some revitalization strategies which will be discussed in this chapter.

Making Dayton an Immigrant Friendly City

In the United States, numerous cities have experienced urban decline; however, some have developed policies to regenerate the dying central city. Many of these cities are actively combating the decline by launching creative initiatives aimed at addressing vacant and abandoned properties. Declining cities in the Midwest have sought to revive their economy by becoming intentional about how immigrants are perceived and have developed strategies to make their foreign-born "feel at home." These local immigrant-friendly policies are situated in a broader political context both at the state and federal level. However, these policies are accepted differently at the state level. For instance, the Indiana state government has taken a conservative standpoint on these policies which is obvious in their low support for local councils to deliver services like the ESL program in public schools (Shepard 2016). Ohio, on the other hand, has provided support for immigrant-friendly policies like the Welcome Dayton plan, which stands out among welcoming plans and even got federal recognition during Obama's regime when immigration policies were at ease.

My interview with Ruiz, a member of the Welcome Dayton committee, revealed that the idea of Welcome Dayton was birthed as a result of discrimination against the immigrant community in Dayton. This implies that Dayton has always been an immigrant destination for secondary immigrants even before the plan came to life. However, the majority of foreign-born were being discriminated against when it comes to housing and employment opportunities. To corroborate this, Katherine, a member of Welcome Dayton, explained that the HRC board was particularly concerned after noticing discrimination against primarily the Latino community and the African refugee community. The need for this initiative became important after realizing that individuals are being discriminated against and yet are not coming to seek justice and to file a complaint and have it investigated. Shepard (2016) identified the consequences of this unfriendly environment to be a further out-migration from the already declining population. To avoid this in Dayton, the City manager and commissioner then met to discuss how Dayton could be more welcoming. Katherine summed up the need for this initiative when she said:

> "The city manager was interested because we had this large increase in
> immigrants in our community and it was very noticeable that people were starting
> small businesses and buying homes. There are kind of two ends of the spectrum
> happening at the same time where, you know, we have people coming to our

community and having faced population and economic decline for decades, to
have a new growth is incredibly important. And yet those people that are coming
are being discriminated against. So, we really ought to think about how we rectify
that, how do we create a welcoming and inclusive community so that those that
are here are able to live up to their fullest potentials."

Ruiz also mentioned that the majority of them weren't just scared of how they would be
perceived by society; some were scared of deportation and so they avoided anything that
had to do with government officials which explains why they remained silent even when
they were discriminated against. It is therefore safe to say that Welcome Dayton was
needed to help immigrants come out of their "shell" and integrate with the community.
This was further explained by Katherine when she said:

At first, we noticed immigrant community were secluded and did not contribute to
the different discourse as at then, even when they were publicly invited to
programs by the HRC they wouldn't show up and at the same time they weren't
reporting cases of discrimination which we all knew they were experiencing.
To corroborate that there was indeed discrimination against the immigrant

community, Dovian, a secondary immigrant who had lived in Dayton for over 10 years,
explained in detail his experience when he first got into Dayton. He explained the
discrimination in Dayton compared to Baltimore where he had lived for five years and
had already become a US citizen before moving to Dayton. However, this didn't stop
people from looking at him differently and hindering him from investing in the city. He
was particular about how difficult it was for him to acquire a space to start his trucking
business and how he was made to undergo what he termed "unnecessary" verifications
before his business was approved. He explained that despite buying an abandoned
property for his business which in his words "would help grow the economy of the city

through property tax which has been unpaid for years" the officials still had to make him

fulfill several requests which didn't only eat deep into his pocket but also wasted his

time. It wasn't until he got angry with the approval board before they allowed him to start

his business. In his words, he told them bluntly "I have a feeling you don't want me here,

you want me to do this, do that and you all should understand it's only my brothers and I.

We are yet to make any profit so it is difficult to fulfill some of these requirements." To

conclude that there was indeed any form of discrimination, I probed further if other

native business owners go through the same ordeal before getting approval. His response

shows that he wasn't aware of the requirements native residents have to meet. However,

he hinted that they might not be troubled as much as he was when he said:

> Before I embarked on the project, I spoke with a business owner here who is a
> native resident to understand what I will need to do before getting my approval
> and everything he said was quite different from what I was requested to do. He
> didn't have to go through what I went through. This is why I told them I had a
> feeling they didn't want me here in the first place or maybe they didn't want me
> to invest here.

To further explain the need for this initiative, one major challenge of immigrants

was the problem of communication. The majority of them came from non-English

speaking countries so it was difficult for them to express themselves upon arrival in

Dayton. This was not a recurrent issue in my interviews with policymakers. However, it

surfaced in conversations with immigrants most especially the secondary immigrants

(Those that settled in a place before relocating to Dayton). Dovian who had lived over ten

years in the city mentioned that Dayton did not provide translation services as at the time

he migrated from Baltimore where he enjoyed these services. He continued that even if they did have these translating services at his time of arrival in Dayton, they weren't made known to him.

All my findings on why Welcome Dayton was important revealed that it wasn't aimed to be a revitalization plan for the city or attract more immigrants at least when it was being formulated. Instead, it was intended to address the discrimination of immigrants and the need to integrate them into society to explore their full potential and make them feel at home so they don't relocate to other neighboring cities. This was supported by Ruiz, a policymaker when asked why the city decided to apply the friendly immigrant strategy to revitalization he stated:

> That wasn't a goal of welcome Dayton, it was a nice side effect, we didn't talk about that when we were putting Welcome Dayton in place. We were just discussing how badly refugees and immigrants were being treated so we wanted to try to make Dayton a better working place for them and that turned out to be great side effects. The reason we want them to feel more at home is to encourage them to use their talents and abilities, so it benefits them and the city too. We are trying to make people take part in public life not having to hide, not having to feel afraid. what we are trying to achieve is to make sure that we are giving everyone the opportunity they deserve to live.

From this statement, I argue that even if Dayton had been a burgeoning city, there is still a possibility that this community-driven political framework would exist if there were discrimination against the immigrant community. Although, it might not have been as involved as it is now. This proves that this initiative was born out of compassion and social responsibility of the policymakers and the economic redevelopment was just icing on the cake for the city.

Formation Process of Welcome Dayton Initiative

After establishing the need for this initiative, it is important to also discuss how it was formulated. From my conversation with Ruiz and Linus who were actively involved in the formation process, it was discovered that this policy adopted the community based approach where all stakeholders from different groups were represented. After seeing the struggles of immigrants in the community, the commissioner met with the then-city mayor and they both met with the chairman of the Human Relation Council which is known as the "conscience of the city." Together these three people called twenty individuals from different sectors and held several meetings, where they talked about what they would like the community to be, what sort of attitudes, what sort of practical effects, and what they would need to do to make the city more welcoming.

According to the Welcome Dayton report (2011), the director of the HRC then came up with a strategy to go public by inviting people who had a say—people who are actively engaged in supporting the integration of new residents. He publicized an invitation letter that explained the benefit of becoming an immigrant-friendly city (see appendix 1). Welcome Dayton was then created through a series of community conversations where they discussed the possible impacts this integration might have on the city. This is important in the discussion of policy formation because it allowed a lot of community buy-in where native residents had a say. Katherine, a member of the welcome Dayton committee, explained that at these round table discussions, Tom Wahlrab- the former executive director of the Human Relations Council, raised questions which aimed to address the fears of native residents in becoming immigrant friendly. These questions were important because if people don't have that opportunity to voice their doubts and fears, and to be able to have them heard, then they might be more resistant to this initiative. To know if there was bias in the selection of the people involved in this process, I enquired from Katherine how participants were selected in these open discussions and she explained that the invitation was opened to everyone interested, and they used a snowball method to further get people involved in the conversation. In her words: "we encouraged people who attend to invite people who they think would be interested in these types of conversations and this included some immigrant as well."

The Welcome Dayton plan emerged from these conversations to provide a framework for communication, implementation, and partnerships among the municipal government, local organizations, businesses, and community members. To efficiently reach its goal, four sub-committees were formed—business and economic development, health and social services, local government and justice system, and lastly, community, culture, arts, and education. Each sub-committee then came up with its plans to help admit immigrants easily into the society.

I argue that the policymakers should be applauded for their doggedness in the pursuit of this initiative. Not only did they do a thorough job bringing people together and ensuring those that were involved (immigrants and native residents) had a say in the formation process, they also remained faithful to their dreams even when the immigration policy at the federal level was in opposition to their goal. In the words of Ruiz: "I'm proud of this city and proud of my colleagues because you know politically back when this was passed, immigration was a tough topic for all politicians to talk about, it's easy for me like I said I'm a grandson of an immigrant so it's the life I live it's the water I swim in but for my colleagues, they passed this plan and they agreed to it without their soul connection they knew it was the right thing to do, so I am really proud of them."

Framework of the Welcome Dayton Initiative

From the Welcome Dayton report (2011) and my conversation with Linus and other policymakers, it was revealed that Welcome Dayton stems from the several recommendations contributed by each sub-committee during the formation process. The success of this initiative depends solely on how each sub-committee handles its responsibilities. As I mentioned in the previous section, the framework of the initiative is covered by each sub-committee which has a distinct role to play in ensuring the city remains immigrant friendly. The goals and recommendation of each sub-committee are discussed below:

First, there is the Business and Economic Development sub-committee whose objective is to create a channel that would encourage the growth of immigrant businesses in Dayton and also create immigrant business clusters to achieve this. They set up two main goals which would help the city create programs or policies that would:

1. Rejuvenate the community by investing in a geography with immigrant businesses who are more willing to populate the city area.
2. Help ease burdens for anyone specifically immigrants who want to open new businesses serving whomever, wherever (Welcome Dayton 2011).

In the Welcome Dayton report, this team gave recommendations on how the city could encourage immigrants to start-up businesses which would have a positive impact on both residents and the environment. One recommendation that stood out from this report is the encouragement of entrepreneurship growth along East Third street (see figures 3 & 4).

Figure 3

One of the Financial Institutions on East Third Street (Google Earth 2021).

Figure 4

One of the Grocery Stores on East Third Street (Photo: Adeuga, M)

They identified this area as a perfect location for business start-up hence the city

government was advised to pay adequate attention to this area and suggested that they

should encourage an easy transition for potential investors in these areas. The location

was chosen strategically because of the factors which economically are considered

"factors of production." First, this committee identified that East Third Street enjoys a

location between downtown and Wright Patterson Air Force Base which houses over

30,000 residents. Thus, it is an area easily accessible to a large customer base. Also, this area is known for its continuous growth of immigrants which implies that more immigrants would likely want to invest in an area where they believe they would enjoy the same cultural value as their neighbors. The area also offers abundant commercial space and proximity to academic, religious, and financial institutions (figure 5) which automatically makes it a preferred location for immigrant investment. To support entrepreneurship growth in this node, the business committee provided some support services suggestions to the city, including providing grants for entrepreneurs, as well as what they called "retail incubator" opportunities which would boost entrepreneurial growth and perhaps serve as a foundation for new businesses in this area. Secondly, they recommended the creation of an inclusive community-wide campaign around immigrant entrepreneurship that facilitates startup businesses, opens global markets, and restores life to Dayton neighborhoods. To achieve this, they suggested that the city should:

1. Facilitate, coordinate, and host the efforts of existing business development teams to become more immigrant-friendly in their program offerings rather than create a new entity.
2. Messages, material, and presentations should provide information about the benefits of immigrants' entrepreneurs, awareness of how their needs might differ from other entrepreneurs, existing resources available to help address these issues, and basic cross-cultural etiquette.
3. Focus on Dayton's history of innovation and its immigrant past to overcome fear and embrace the full richness of cultural diversity.
4. Work with immigrant social networks to nurture the rebirth of strategic Dayton neighborhoods (Welcome Dayton 2011).

Next is the Local Government and Justice System. The goal of this sub-committee is to improve the language interpreter capabilities and to increase immigrant participation in government, community organizations, and activities; increase trust and communication between immigrant communities and law enforcement; and overcome language barriers in the court system and prosecutors offices. They provided recommendations to

1. Promote increased access to government services for Dayton's residents who are limited English proficient (LEP) by having language services available.
2. Adopt law enforcement policies that are immigrant-friendly throughout the greater Dayton area.
3. Increase involvement of immigrants in policy making and community programs by removing barriers to participation and encouraging civic activities.
4. Ensure access to the justice system for immigrants, regardless of language barriers or status.
5. Implement a municipal identification card program for community residents who are not eligible for any other accepted identifying document.
6. Educate immigrants about government services, laws, and social services and educate social service providers and government officials about immigrants (Welcome Dayton 2011).

The third sub-committee is Social and Health Services. These committee members focused on the accessibility of immigrants to social services available to them. Their goal was to remove every form of language barrier hindering immigrants' access to social and community services. Some of the barriers include insufficient translated resource information and lack of proficient English interpreters. They recommended that the city should build a website that would provide immigrants a list of existing health and social services which are available to them. Not only this, but they also recommended that immigrants should be educated about government services and that government and service providers should be educated on the needs of immigrants. They also suggested that interpreter volunteers could be encouraged by creating training to make them efficient. In the executive summary, they encouraged partnerships with private organizations that would like to work with immigrants using the existing framework.

Lastly, there is Community, Culture, Arts and Education. This committee formed three goals which prescribe education for non-English speaking immigrants, encouraging youth in community building, and putting the cultural diversity of Dayton to good use. They gave several recommendations that the city should adopt to achieve these goals some of which include the training programs that help workers understand the cultural barriers experienced by immigrants. This is important because the majority of the immigrants I was able to interview mentioned cultural barriers which hinder them from fully exploring their potential in the city. Also, they suggested building a base of ESL and literacy tutors to volunteer in existing/expanding programs. To encourage cultural diversity integration, they recommended a soccer event where participants would represent the different ethnic/cultural groups in Dayton.

Implementation of Welcome Dayton Initiative

In its implementation stage, the city commission passed an order, which nominated a Welcome Dayton Committee to oversee the smooth running of this initiative. This committee meets on a semi-monthly basis to discuss the progress of the initiatives while receiving reports from each sub-committee which must have met during the off months. Secondly, the relationship between the Human Relations Council and the city manager led to the establishment of a part-time office that houses this committee and to "facilitate and coordinate the efforts of community organizations and businesses since most of the recommendations are not within the mission of the city itself" (Welcome Dayton Report 2011, 12). Thirdly, the city manager is assigned the responsibility of managing the affairs of the plan which cuts across the mandate of the city. Lastly, the city commission encouraged "immigrant groups, other government agencies, community institutions, and the business leadership to undertake their initiatives, beyond this plan." It is believed that this would not only make Dayton an immigrant-friendly city but also shape Dayton into a commercial center when immigrant groups incorporate their ideas into the plan (Welcome Dayton Report 2011).

After the creation of the goals and recommendations by each sub-committee, and the three phases of implementation by the city manager, they decided to further divide each committee into a smaller group which would help individual organizations concentrate more on specific roles within the context of their parent committee goals. In the words of Linus from the Chamber of Commerce:

> We decided to divide up into smaller subcommittees so we can advance some of the work that is kind of more specific to individual organizations. So, we have a health care committee that focuses on how to make our health care system work better for our immigrant communities. I work with businesses and small entrepreneurs. And so, as we sort of look at our subcommittee work, that kind of helps us delve down a little deeper. And then we meet as a large committee every quarter and sort of a report on the individual work of the sub-committee.

This is commendable because not only does it allow each sub-committee to reach out to a large number of immigrants it also makes them more accessible through the various programs they create. The Chamber of Commerce, for example, organizes training sessions for business owners on how to hire new immigrants and what to expect from them. Dovian, an immigrant business owner, mentioned how one of these programs helped his business in the employment process of another immigrant who didn't have a social security number. According to him, it was at one of these programs that he raised this issue and the moderator told him to link the worker with another immigrant support group which eventually helped him get a tax identification number.

Aside from the many initiatives that make up Welcome Dayton, several other grassroots initiatives outside the framework of Welcome Dayton have sprung up. However, they still support the goal of Welcome Dayton, and fortunately, they have been encouraged by Welcome Dayton. One of the initiatives mentioned by Katherine during my interview is the Natural Helper which serves as the intermediary between the immigrant communities and Welcome Dayton initiatives. This group was formed by people who volunteered to help Dayton become a better place for immigrants. They are the people immigrants turn to when they need help navigating the system. For example, they would recommend where immigrants can get ESL lessons. A member of HRC referred to them as the "front door" for immigrants to get integrated into the city. This group of people are known by the Welcome Dayton committee/staff and they provide monetary and financial support for them. This was supported by Linus, a member of the Welcome Dayton Committee, when she said:

We're looking at how can we provide stronger support so that those individuals
don't get burnt out and also, how do we grow that capacity and sort of Welcome
Dayton natural helper program and are in the process of that because that, like, the
ideal long-term vision is to have a welcome center. But with that being resource-
intensive, how are you able to create more of a network or a backbone to that kind
of concept, because they are the individuals that are the front door, right. It's
people in the community much before they would go to a government agency to
seek services.

This initiative has indeed helped immigrants fulfill their dream of living in a welcoming

city. This was deduced from my conversation with Yun, an immigrant who has lived for

2 years in Dayton, who attested to having experienced support from a group without even

realizing its affiliation with Welcome Dayton.

Several other institutions have joined forces with Welcome Dayton to help build a

welcoming city one of which is the law enforcement agency. The Dayton Police

Department was frequently mentioned both in my conversations with policymakers and

in the Welcome Dayton committee meetings. From my findings, it is obvious that the

police department to a large extent has contributed to a welcoming atmosphere. I inquired

how they have been doing this from Ruiz who explained that most times, police officers

are lenient with immigrants in minor cases such as traffic offenses and they wouldn't

even ask immigrants for their legal status in the city. This was corroborated, in one of the

Welcome Dayton committee meetings where it was mentioned that the police department

had a policy that ICE would not be contacted for low-level offenses. They successfully

did this for nine months straight in 2018. I inquired how the police department has been

able to accomplish this and how Welcome Dayton has supported them, and it was revealed by Ruiz that Welcome Dayton provides training on what to expect when dealing with immigrants and also provided interpreters for the department when the need arises.

Welcome Dayton as an initiative also relies on partnerships with private organizations to reach its goal. Each sub-committee has been able to partner with different agencies. Some of the paramount partnerships that came up in my interview and analysis of committee meeting minutes include the Catholic Social Services of Miami Valley. Welcome Dayton works closely with this organization that is responsible for the resettlement of refugees and their families. This organization is the first contact for many refugees. The first thing they do once they receive refugees is to provide a temporary housing unit for them after which they orient them on the cultural difference and what to expect in Dayton. During this process, they identify those who are illiterates and non-English speakers and help them find the most suitable English for Speakers of Other Languages (ESOL) classes. This organization also works with local employers and can match refugees' skills to the needs of employers. Lastly, they act as an intermediary between refugees and the community resources that are accessible to them.

My conversation with Linus, a committee member, revealed how Welcome Dayton has partnered with both national and municipal organizations. When asked to give some examples of these organization and what they have achieved, she responded that:

> I've worked closely with what is called the New American Economy. And they have helped us produce an economic report that shows the impact of immigrants and refugees on our nine-county region. And so, they are we're able to do sort of all the economic research and pull together the data so we could release that

report to the community. The other sort of project that we've worked on is called the Global Dayton report. And this was led by Montgomery County and some other partners.

This said project has served as an evaluation tool for Welcome Dayton and has also shown how immigrants have impacted Dayton. The partnership with different organizations for evaluation purposes seems to be a re-occurring process because, in one of the Welcome Dayton Committee meetings, a member mentioned how they have decided to collaborate with Dr. Suban (a professor from Wright State University) to research how immigrants in the Dayton area are doing. This would allow the Welcome Dayton Committee and staff to measure progress in specific areas and allow the WD Coordinator to audit their progress regularly. And then more recently, they have focused on working with the University of Dayton and Wright State University to develop innovative ideas on how international students can be more integrated into the community by providing them internship opportunities and encouraging them to stay in Dayton and work after the completion of their degree.

All these have supported the fact that Welcome Dayton does not operate on its own but rather relies on several initiatives and partnerships to succeed. The plan uses a systematic approach where immigrants don't even need to be in contact with Welcome Dayton before accessing their services. Linus explained that there are multiple entry points to access the resources of Welcome Dayton. For instance, an immigrant could visit the Chamber of Commerce if they need advice on business startup. Likewise, non-English immigrants could request the service of a translator if they visit any community center like the hospital. In Linus' words "We wanted them eventually to become established through what welcome was able to offer."

The systematic approach of Welcome Dayton explains why Yun, an immigrant, claimed not to know anything about Welcome Dayton despite its national recognition. This was corroborated by Katherine and Ruiz who explained that this is not surprising because most of them do not have direct contact with the Welcome Dayton office. Rather, they are being attended to either by partners of Welcome Dayton or some of the grassroots initiatives. Ruiz also stressed that it does not matter if they know the program. All they are concerned about is their well-being and if the city has been welcoming enough. When I asked if individual programs have truly achieved the goal of Welcome Dayton, he responded in the affirmative:

> Generally, the programs have been effective in doing what we want which is to bring people into the light, give people skills make them feel less scared about living here, and then like I said letting them use their talents and abilities for themselves and their families and their community.

Challenges of the Welcome Dayton Initiative

The Welcome Dayton program, despite its contribution, has experienced several challenges which have drawn the plan backward and have created fear within the immigrant community. However, the Welcome Dayton committee has tried to strike a balance in the city by introducing several measures that would not put their immigrant community at risk. In this section, I will discuss the main challenges that came up in my interviews and document analysis.

Despite the tremendous work put in to become immigrant-friendly, Dayton has had its fair share of the unfriendly immigration policy by the federal government. All the policymakers that I interviewed mentioned how the city has experienced a drastic change in immigration patterns due to Trump's immigration policies which discouraged immigration. One of the policymakers mentioned that the last four years had a huge impact on what they stand for because it had the effect of pushing people back to the shadows. This is also similar to what Linus said when I asked about some of the challenges facing Welcome Dayton:

> The challenge, I think, with being a welcoming city and tackling some of the immigration issues as a whole is that the national conversation and national policies have been so tough, particularly the last four years. It's hard sometimes to be able to move forward with our work when there are so many other, legal issues that are coming up for people. And, you know, members of our community are struggling too, for example, are our refugee resettlement services ended up being able to take significantly fewer refugees over the last two years and they had planned and that that impacts the work that they do. So, kind of dealing with that political side of it has been tough.

This implies that several immigrants were having legal issues and as the commissioner mentioned this got immigrants scared and they didn't want to have anything to do with local government officials for fear of being ratted out and this made them reclusive.

These tough national policies have surfaced in different forms in Dayton and several of these instances were mentioned in the Welcome Dayton meetings where committee members expressed their concerns and provided possible approaches to cushion the negative effect of these unfriendly policies. In one of the committee quarterly meetings of Welcome Dayton, Frederick (pseudonym) informed other members that the supreme court of Ohio released a memo that Immigration and Customs Enforcement (ICE) should not be prevented from entering courthouses for immigration enforcement purposes. This particular meeting discussed cases where hospitals allowed ICE to enter and detain patients even when they were receiving treatments. However, another member raised a motion to create awareness for hospitals that ICE only has the right onto their premises if they have a warrant. In another Welcome Dayton meeting, Gabrielle (pseudonym) mentioned another case where ICE detained a woman who had been in an abusive marriage for twenty-five years. When she gathered the strength to report the situation her husband called ICE and got her detained.

All of these immigrants' negative experiences are why immigrants have chosen to remain low-key, so instead of reporting situations to government officials, they chose to live with their situation. To resolve this, several departments and committees strategized to help immigrants come out of their shells. One is the police department that noticed this and being an immigrant-friendly institution, they decided not to involve ICE when minor cases were reported. Also, when Ruiz, a member of the Welcome Dayton committee, was asked how they reacted to this situation, he responded that:

> I think that the less that I or any other city official showed up for things the better we let our immigrants like immigrants resource specialist does reach out to local communities, we let them do our talking for us, you know when they show up people are intimidated or scared, so we were able to continue doing things because they were embedded in those communities and they knew we were complete and they have built up trust, so we try to keep the federal government out of anything everything that we do and let our immigrant resource people do their work in spreading the news.

As Ruiz mentioned, immigrants were scared of contacting government offices for legal counsel due to the fear of being reported to the ICE. This to an extent changed the direction of Welcome Dayton because they had to partner with private legal institutions that could help their immigrants navigate their legal challenges. When I asked how Welcome Dayton has been able to meet its goal of integration during this time, Linus' response showed that these unfriendly policies made them reduce their work on community building, and instead they have placed more attention on legal support for immigrants through partnership. The partnership Linus mentioned was evident in one Welcome Dayton meeting where it was mentioned that attorney Buzz Portune had donated 960 billable hours in a single year to help refugees complete their green card adjustment application. In a single year, he was able to get 60 green card applications approved.

The unfriendly immigration policies at the federal level have also encouraged the growth of groups who are against cities becoming immigrant friendly. One of these groups is "We Build the Wall." This group was formed in solidarity with Trump's border wall project which would reduce the intake of immigrants in the United States. This group has had several town hall meetings in Ohio and the Midwest where they believe there is a "weak wall." Their growth has threatened what Welcome Dayton stands for as they publicize the negative impact of integrating immigrants into US cities. Gockowski (2019) revealed that in one of their meetings in Ohio, a woman explained how immigrants were responsible for the death of her daughter who died as a result of a drug overdose supplied by an immigrant. Some others believed the influx of immigrants in cities is responsible for the employers paying low wages because the labor force has been saturated.

The policies at the federal level have also influenced Dayton's residents' perception of welcoming immigrants. Initially, there was not any opposition to Welcome Dayton, most especially during the formation process. Ruiz mentioned that it was quite surprising to them that no one voted against the plan during the formation process. In his words:

> The only people who came and spoke against it were people from out of town who had no say. We had somebody come from Cleveland we had somebody come from somewhere else and that was it and almost all the emails that we got against it came from different places across the country places outside the county.

This proves that as at the time of formation in 2011 residents were in support of immigrants' integration. When asked how they have been able to reaffirm the importance of immigrants, Ruiz mentioned that they have a community dialogue series called "Voices." Immigrants are given the opportunity to share their experience on their journey, how Welcome Dayton has helped them, and how they have in return helped shape the city. This program was organized because it is believed that hearing people's personal experiences is an effective way of changing minds.

Dayton's Immigrants

Even though the idea behind the formation of the Welcome Dayton initiative was not for revitalization, this plan has successfully attracted immigrants and refugees into the city and has helped rebuild the economic and environmental structure of Dayton. This section discusses the immigrants in Dayton, how they have reshaped the city, and how residents have allowed them to explore their potential. Due to the pandemic situation during this research, I was able to speak with just three immigrants. This implies that the experiences of these immigrant participants might not represent the experiences of all immigrants in the city. Nevertheless, their responses provided an insight into the impact they have had on the city and some of the challenges they experience living in Dayton.

From my interviews and document analysis, it was revealed that Dayton has welcomed immigrants from different ethnic and racial backgrounds. The major group that came up in my analysis were the Latino community, the African community, and the Aishka Turkish community. Not only have these foreign groups increased the population of Dayton, but they have also contributed to the economic redevelopment of the city.

The Aishka Turkish Community

This immigrant community seems to be the most significant because their impact was mentioned by every interviewee in this study. From my findings, the majority of them came into the United States as refugees when they weren't treated well in Russia. A background to their history revealed their migration pattern from when the Ottoman empire collapsed in 1829, their evacuation to Russia, and how they eventually got to the United States. While in Russia, they were treated poorly by the government, experienced police brutality in different forms, and weren't allowed to live freely. Dovian, a Turkish immigrant who has lived over 10 years in Dayton, explained this better when he said:

> So, we were in Russia, we lived there for 20 years, no citizenship most people don't have driver's license, no form of identification, nothing! We worked on farms and we were not allowed to go back to our homeland because we didn't have any documentation.

After several years, the International Organization for Migration (IOM) heard about them and encouraged them to come to the United States promising them a green card that would allow them to work, and if they were law-abiding, they could become US citizens. Upon their arrival into the United States, they could work and move freely within the country.

When I asked immigrant participants why they chose Dayton of all cities in the United States, their responses showed the influence of the city's decision to become immigrant friendly. Dovian explained that choosing Dayton above all other cities wasn't a tough choice for him given his bad immigrant experiences in Russia he wanted a city that wouldn't just welcome him but also allow him to live a peaceful life—free from police harassment and social vices. The welcoming nature of Dayton did not just play a role in resettling immigrants. It has also helped them identify their potential and build a strong network connection which could also be attributed to why the Turkish population has continued to increase in Dayton. This was revealed in my conversation with Yun, a Turkish immigrant who has lived in Dayton for over three years, who mentioned how close-knit their community is and gave examples of how they help one another when the need arises. This supports Portes and Rubaut (2006) that indeed, the existing immigrant community plays a huge role in immigration patterns. The Turkish community has been able to attract more population because of the unity and welcoming environment they have created. They recently came together to build their first mosque since they arrived in the United States and also went further to construct a community center for events and Arabic lessons. It was deduced from the migration story of Yun that these two facilities are also instrumental in the integration of new Turks into the city. Yun explained how he knew no one in Dayton before relocating and the first job he got was to teach children Arabic in this community center. He also mentioned that some new immigrants have been employed to work in these facilities for maintenance purposes while serving as a temporary shelter for some. Dovian, a secondary immigrant, also described how welcoming Dayton is and how the Turkish community has supported him in a way to be a better person in society. He further described how he has encouraged his friends to relocate to Dayton and in his words "two of them are living here now!" Immigrant communities advising people to relocate to Dayton is what Linus termed "self-advertisement" and is one of the objectives of Welcome Dayton. She explained this in my interview when I asked her why she thinks people are relocating to Dayton.

> I think it has to do with the cost of living in Dayton which is really cheap, I mean we have a lot of unused properties which goes for a cheap price. Also, people move to Dayton because of their friends and families who are here and have experienced what Dayton has to offer. You know this is actually a good thing because we don't even have to go out there to advertise that Dayton is welcoming. We just continue our work to become more immigrant friendly and our immigrants will do the marketing by telling their friends to move to this welcoming city.

Interviewees also suggest that being immigrant-friendly can help rebuild a declining city most especially when these said immigrants have experienced an unfriendly environment elsewhere and are in search of better living conditions where they can fully benefit from the available resources and less hostility.

Figure 5

Dayton's Turkish Community Center (Photo: Adeuga, A)

Another pull factor that Dayton has enjoyed over the years is the current state of

the city. Being a declining city with a lot of abandoned properties has opened the door for

re-investment Coupled with its immigrant-friendly policies and investment opportunities,

several Turks have resettled here because of its low standard of living. This was

confirmed by two immigrants who claimed to have enjoyed the low cost of property

acquisition in Dayton. Dovian, who is in the trucking business, explained how he acquired his office space at a cheap rate compared to other cities where he had lived. In his words:

> I bought this place for eighty thousand dollars, there was not a warehouse, obviously the inside wasn't like this, this place was abandoned for 8 years, there was grass and taxes weren't paid for ten years.

If this said warehouse were to be in a wealthier city, it would most likely cost hundreds of thousands of US dollars. However, Dayton's abandoned buildings and unpaid property taxes have made such properties readily available at a cheap rate to immigrants who are interested in buying them. Yun explained how one of his friends buys the cheap abandoned buildings, renovates them, and resells them to make a profit. This again supports the fact that there are many economic opportunities available to immigrants in Dayton and to a large extent, they have taken advantage of these. Some other opportunities that have attracted immigrants to this city is the presence of educational institutions and job opportunities. Both factors have been identified as significant pull factors in migration by scholars like Sahatcija et al. (2020) who suggested that job opportunities and quality of education are important factors that influence migration amongst students. In my interview with Yun, who is also a student, he explained that his decision to migrate to Dayton was influenced by the presence and availability of educational institutions and job opportunities that his friend had told him about. He said:

> I have always wanted a city where I can develop myself academically and at the same work, even if it's a small paying job, I wouldn't mind, I just want to make sure my family does not suffer while I am learning.

To corroborate what Linus had told me on how Welcome Dayton Partnership helps students secure internships, I asked Yun if he was able to secure a job while schooling and how Welcome Dayton has helped. His response revealed that he had a job, though a menial one, and he intends to get something better in the future. However, to the question of how Welcome Dayton has helped get jobs, Yun responded:

> No, I didn't go through Welcome Dayton, a friend of mine mentioned this place to me and I went there, got an interview the same day, and got the job, I did not even have to submit any of my credentials well, it menial job so they didn't require that.

I believe it is quite understandable that the Welcome Dayton does not have to provide jobs for all immigrants as the initiative may not even be aware of all job openings within the city. However, in his explanation, he mentioned how he had attended a program where they helped him build a resume and gave him job interview tips but he was not sure if this was organized by Welcome Dayton.

Impacts of Immigrants on the City

My conversations with both immigrants and policymakers revealed the several ways Dayton has enjoyed the benefits that come with being immigrant friendly. These immigrants have impacted Dayton socially, environmentally, and economically. First is the population growth which has led to economic benefit. According to a New American Economy Report (2018), the foreign-born population in the Dayton region represents about 4.6 percent of the total population and this increased by 17.6 percent between 2012-2018. This was corroborated by Tetula- A Welcome Dayton committee member when he said: "Our city has experienced a population increase which is a result of the increasing immigrant community."

My conversation with Tetula also revealed that this immigrant population has contributed to the buying power and tax revenue of Dayton. This was supported by the New American Economy Report, which stated that immigrants in the Dayton region earned 2.8 billion dollars out of which over 500 million dollars was paid in federal taxes and 260 million dollars in state taxes. Understanding how they have contributed to the economy, I asked Linus about the type of jobs these immigrants are doing, and she responded that:

> So, there is a lot of research that shows that immigrants are much more likely to be entrepreneurs. So, they are much more likely to be entrepreneurs than sort of the native-born population. So, they will start their own small business, whether it's a restaurant or a lot of times it's a restaurant or it's a hairdresser or beauty business or a small sort of like logistics company, truck driving business warehouse, things like that. So those small entrepreneurs make up a big portion of our immigrant population. And then it's other industries that you would expect. The STEM industries, our science, technology, engineering, and math professions have a high proportion of immigrants, often because they are the most trained in the area. So, our engineering jobs, our sometimes our medical jobs, things like that, those have a higher proportion of immigrants as well.

The New American Economy provided statistical support for this claim in its report, showing the major industries where immigrant entrepreneurs invest. Figure 6 shows that 17 percent invest in professional services while 16.7 percent invest in general services. Interviewees suggest this high investment in professionals is attributed to the educational attainment of immigrants.

Figure 6

Industries where Immigrants Entrepreneurs Invest in Dayton (New American Economy Report, 2020)

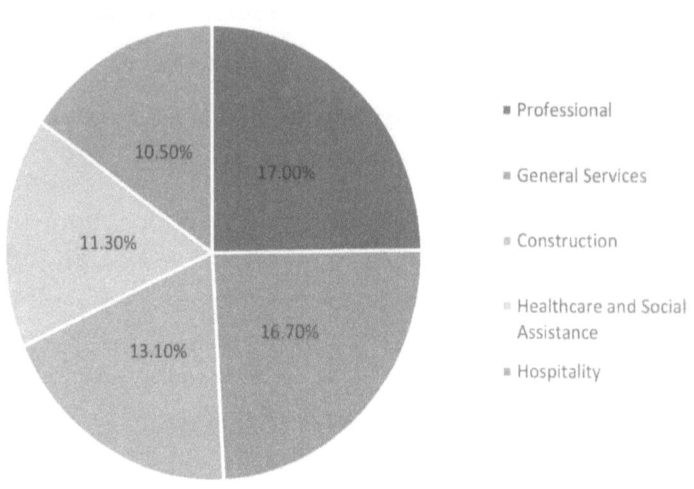

In terms of investment, another sector where immigrants have impacted is the educational sector of the city. My interview with Yun, who has lived for 2 years in Dayton, revealed how he has been determined to obtain another master's degree in the United States despite the cost and struggle he has to endure to achieve his dream of getting a decent job in the city. The New American Economy report revealed that 45.4 percent of the immigrant population who are aged 25 and above have a bachelors' degree while 23.6 percent of them have an advanced degree. Not only have these populations helped fill up institutions' classrooms, but their tuitions have also contributed to the economic growth of the city. The New American Economy Report revealed that these two classes of population have contributed a total of 250 million dollars in fee payments and other educational expenses. The report also indicated how foreign students have helped keep over 3,083 local jobs open by taking up menial jobs which the native-born would not opt for. The importance of this is evident when Yun went for his menial job interview and got employed the same day. This shows the desperate need for the labor force in local businesses which immigrants have been able to meet. Another area of significant immigrant impact is neighborhood renewal. It has been established that a major characteristic of a shrinking city is the presence of abandoned properties and unfortunately, Dayton isn't an exception to this phenomenon. However, immigrants have begun to change the face of this city by investing heavily in abandoned properties either to resell in the future or to inhabit. Ruiz, a, member of the welcome Dayton committee mentioned one of the areas that have benefited from this housing investment when he said:

> If you drive out northeast of downtown if you are at the community center there are a lot of houses that have been fixed up, it was a neighborhood that was on the way to being abandoned. When the visitors came, they bought the houses, they started businesses they fixed everything up.

Also, my conversation with Dovian, an immigrant who has lived for over 10years in Dayton, revealed how he had invested hundreds of thousands of dollars renovating his current office space which used to be an abandoned property. Lamidi, another committee member of Welcome Dayton corroborated this and mentioned how they have worked with both Latino and Turkish communities to revitalize neighborhoods in North Dayton.

He mentioned how Welcome Dayton has helped streamline investment procedures making both Latino and Turkish communities invest easily in property acquisition. They also went further by providing legal advice to potential investors when needed. In the words of Lamidi, a committee member of Welcome Dayton:

> …what we did was work with the Aishka Turkish community and the Latino community to start to revitalize some of the neighborhood. So, a lot of the work that was done earlier on was focused on North Dayton which is where a lot of the Aishka Turkish settled. If you take a drive around there you would see how it has been reshaped.

Neighborhood renewal and economic investment by immigrants, despite being a benefit have led to uneven development between the two main regions of the city. After taking a drive around the city, I noticed how the western part of the city had a lot of abandoned buildings (see figure 7) while the northern part (see figure 8) seems to be enjoying the better part of the redevelopment. When I asked the policymakers why this phenomenon is evident in the city, they all attributed it to uneven investment by immigrants. Linus explained this further when she said:

> You have identified a real challenge that our city has. There is a lot of disparity between where our overall community investment has been, either on the west side or on the east side. I think for our Turkish population, part of why they were investing in the Northern part is because they had some initial investments and wanted to grow around what is familiar to them. I mean there is a lot of cheap property across Dayton like there is a lot of commercial buildings that you can invest in cheaply and that's not limited by geography.

This supports Portes and Rubaut's (2006) argument that immigrants would most times invest or relocate to areas where they already have a growing cultural community. It is therefore important that shrinking cities should ensure the first set of immigrants are made comfortable in their environment if they intend to adopt the immigrant-friendly approach of revitalization because oftentimes these people are used as models for future immigrants.

Figure 7

Abandoned Building in a Street in the Western Part of Dayton (Photo: Adeuga, M)

Figure 8

A Renewed Neighborhood in the Northern Part of Dayton (Photo: Adeuga, M).

The concentration of investment and redevelopment of neighborhoods in the northern part of the city has led to a surprising special dynamic where some of the Turkish population are now moving to Tipp city—a community outside the jurisdiction of Dayton. Shepard (2016) identified this as a common phenomenon in shrinking cities where the cycle of reinvestment starts to hurt the city. Dayton has begun to experience this although it was only mentioned in only one of my interviews. Dovian mentioned how some of his friends have begun to leave Dayton for a neighboring city. In his words:

> The Northern wasn't this developed when we moved here but the Turks have invaded this space, but you know the more this part is developing, the more things are becoming expensive. Now houses that used to be sold between $50,000-

$80,000 are now sold for almost $150,000. So, people are now moving to Tipp
city which is about 20 minutes away from here.

I probed him further to ascertain why the sudden increase in housing cost and why people
are migrating to Tipp City instead of the western or any other part of Dayton. His
response revealed that the sudden increase in housing cost could be attributed solely to
the continuous increase in population which speculators have taken advantage of. Also,
he mentioned how the northern part is beginning to deteriorate gradually and some
wealthy ones who could afford better houses elsewhere are beginning to move. He
explained this by saying:

Michael, you know this place used to be very beautiful, I mean, it still is but
people have found a better environment. You were asking why people wouldn't
want to move to other parts of Dayton, sincerely what other part is better than this
area in Dayton? I mean in terms of where Turks live. Actually, there is no better
place for us than the North, and talking about Tipp city the Turks community is
beginning to grow there as well and the houses are also affordable maybe around
200,000 and it's a new neighborhood.

Another impact that was identified is the cultural diversity that immigrants bring
into the community. Dayton has enjoyed the presence of a diversified immigrant
community from the Asians to the Polynesians. Figure 9 shows the different regions of
origin of Dayton's immigrants. One of the benefits of this diversity was mentioned in my
interview with Katherine when she explained how native residents have learned different
cultural values through the various events hosted by immigrants. In her words:

There are several cultural events organized by the immigrant community in
Dayton, at times they come to us to help them publicize their event and we have

seen how some residents show up at this event and some even go to the extent of dressing like their host.

This points to the fact that some native residents have integrated with the immigrant community which proves that the majority of them are welcoming. Aside from the cultural exchange between these two communities, I identified the cross-cultural learning opportunities that the presence of immigrants has caused. In my interview with Linus, a member of Welcome Dayton committee, he explained how the Welcome Dayton partnership has been able to come up with a mentoring program for native residents. These programs provide opportunities for native-born residents in the city to identify how they can work closely with immigrants. She gave an example of what I term a symbiotic learning process when she said:

> We work with the native-born populations and helping the native-born in the Dayton community identify opportunities to work more closely with immigrants and refugees. We bring the native population together with our immigrant population so they can learn from each other, sometimes with mentoring programs. So, in the case of businesses, we may partner a small business owner who is native, born from Dayton with an immigrant business owner so they can learn from each other.

Lidia (2018) identified this type of relationship as one of the benefits of cultural diversity which encourages access to a diversity of products. Here, cross-cultural learning gives rise to a diversity of goods produced or services rendered since both parties are allowed to understand the need of their community better.

Figure 9

Diversity of Immigrant Community in Dayton (Shrider, 2017)

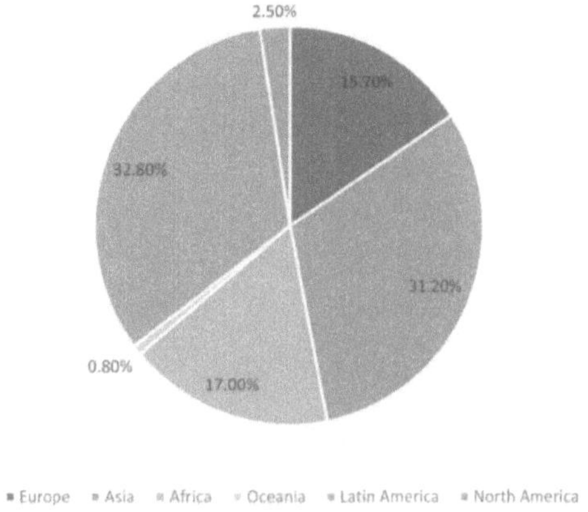

Struggles of Immigrants in Dayton

Studies on immigrant experiences have identified how new immigrants struggle in cities, most especially when it comes to settling down—getting accommodation, employment, and locating important places (Shepard 2016). Although Dayton being a welcoming city has tried to mediate these challenges by partnering with the refugee resettlement program and other private organizations, new immigrants like Yun still experience difficulties. My interview with Yun, a Turkish immigrant who has only lived in Dayton for 2 years revealed his struggles since he migrated to Dayton.

First, Yun- who is also a master's student- mentioned how he has struggled with the educational system in the United States which he considers too expensive and more stressful when compared to his previous educational experience outside the United States. In his words:

> It is strange here kind of, first I came to Dayton when I heard there are jobs here: then I got here and I became interested in education because I couldn't get a befitting job and I already had a master's in mathematics from Russia but it was useless here so, I am here starting another masters' degree which is too expensive for me.

From this conversation, the struggle of getting a job with an international credential is evident. In the United States, it is quite hard for immigrants to get their dream job with a foreign degree, and oftentimes they are left with no choice but to enroll in a US program. Yun described how his master's certificate from Russia has been reduced to nothing in Dayton and how he didn't have a choice but to start a menial job to support his family and fund his new educational path. I argue that this phenomenon is responsible for why the number of immigrants enrolling in master's and advanced degrees keeps increasing in Dayton compared to native-born residents.

Figure 10

Education Demographics in Dayton (New American Economy Report,2020).

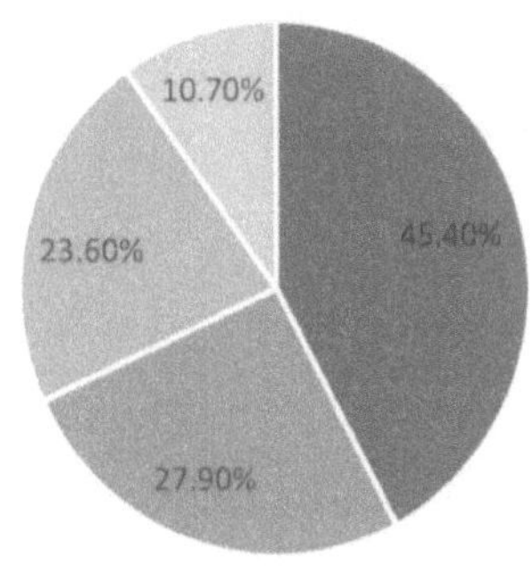

Figure 10 above shows how immigrants seem to cherish education. However, what it didn't reveal is the struggle of immigrants to pay tuition. My conversation with Yun revealed how some of them are struggling to pay their tuition_fees. Yun expressed the pain he has to go through before raising his fees when he said:

> You know fees are too much, I need to work extra hours to raise the school fees and I am not able to spend time with my family or even go on a vacation. All I do is work, and this even affects my schoolwork because I can't submit assignments on time, sometimes I even get to school late.

From this conversation, I deduced how struggling to make ends meet has affected his productivity in school and therefore recommend that the Human Resource Council should encourage jobs to employ some of these educated immigrants. I believe while they are on the job, they can then create time to further develop themselves by taking part-time classes with ease.

Another significant struggle of immigrants is the language barrier. Scholars like Ingo and Otten (2014) have identified one of the obstacles of integration to be the language barrier between the immigrant community and their host. The majority of Dayton immigrants are non-English speakers which put them at a disadvantage when integrating. To combat this, Welcome Dayton has organized ESL classes for non-English speakers. However, my conversation with Yun revealed how his job has hindered him from attending these classes when he said:

> Another problem I had when I first got here is the language, everybody speaks English, but I speak Turkish and I wasn't very good in English although I have improved over time because of my friends who have stayed longer here.

I asked if he attended any of the ESL classes organized by Welcome Dayton, and his response revealed how his job does not make him available for that. In his words: "No I could not attend because I started working not quite long after getting here. So, I decided to self-teach myself and with the help of friends I became better." My conversation with Yun also supported Ingo and Otten's (2014) argument that the language barrier could hinder the economic prospect of a shrinking city especially when immigrants do not speak the same language as their host society. I asked Yun how his inability to speak English affected him economically, specifically in terms of getting jobs, and he responded that it did not because he didn't have direct interaction with customers. However, he mentioned how he was denied a job in the past because of his inability to communicate fluently in English. I therefore recommend that the Human Resource Council should further intensify the publicity of these ESL classes and make new immigrants aware of the importance of such classes.

Despite the benefit that comes with the cultural diversity in Dayton, it has also been a struggle for immigrants to integrate into the new culture they found themselves. Just like Yun, the majority of the Turkish immigrants are Muslims and they have found it quite stressful adjusting to the culture in the United States. Yun mentioned how he finds it difficult to eat some of the foods served in native-born restaurants. In his words, he referred to this type of food as "Hala" which is against their religious beliefs. I probed further what he meant by this, and he explained that as Muslims, they aren't supposed to eat meats prepared by non-Muslims because Muslims have a particular tradition that must be followed when preparing their meat.

Also, the credit system in the United States has limited new immigrants' access to acquire properties in Dayton. Yun who has lived in Dayton for two years explained the financial struggle he had to go through when he arrived in Dayton. He explained how he wasn't able to rent a house because he didn't have a credit history. In his words:

> So, when I arrived in Dayton, I tried to rent a house and the agency asked for my credit history, I told him I didn't have one that I have a cash, then I was advised to go and get a credit card. Upon arriving at the bank, they asked if I have a job, I said no I did not have one yet. Now they are saying I should get a job before applying for a credit card or I should get someone to sign for me. Everything is based on credit and it was just frustrating when I first got here.

I asked how he was able to maneuver this period and he revealed how a member of the Turkish community agreed to stand in for me. This again proves how this community always supports itself when the need arises. One common thing amongst all the struggles identified is that they all seem to be normal challenges for new immigrants irrespective of the city's context of reception. However, I recommend that city officials should inform immigrants on what to expect in the city and orient them on possible ways to alleviate these challenges.

Residents' Perception to Being Immigrant Friendly

Welcome Dayton hasn't received much backlash directly from native-born residents. However, my conversation with the immigrants showed there are two spectrums (welcoming versus un-welcoming) of how residents have perceived them. My interview with Lamidi, an immigrant who has resided in Dayton for over 20 years and is currently a committee member of the welcome Dayton, revealed that he hasn't had any problem with native residents and has enjoyed their welcoming atmosphere. This was evident when he said:

> You see I always tell people I felt Dayton was my home when I got here. Dayton is unique because it's had a history with immigrants, it's always had a unique place in America as a welcoming city. If you go to old north Dayton, you will see the European settlement and the flag of the Turkish. A lot of the welcoming has to do with the fact that the leader and politicians have done tremendously well to welcome. They respect everybody, Imagine, I was recently voted into a position by these same residents.

This statement attributes the welcoming nature of residents to the work that was done during the policy formation when the HRC director listed the benefits of becoming an immigrant-friendly city and allowing residents to voice out their opinion about the plan. From the perspective of Lamidi, residents are welcoming to immigrants. However, the opposite was revealed in one of the committee quarterly meetings where they discussed the case of the immigrants' mosque vandalization. Although the culprits weren't caught at the time of discussion, the committee members believed it was an Islamophobic act. This incident shows that some members of the community had problems with the Islamic group. Dovian, who has lived in Dayton for over 10 years, discussed discrimination the immigrant community faces by some of Dayton's residents. My conversation with him revealed, for example, how his child is treated poorly in school because of his Arabic name. In his words: "This boy was being taunted in school. Just because of his name, he was called a terrorist." It could be argued that he was a young boy and those who called him such names were young as well and they probably do not know how hurtful their name-calling is. However, Dovian's experience further revealed that even the teachers and principal weren't taking action to correct these kids until he visited the school. In his words:

> The kid went to the teacher, the teacher didn't do anything. He went to the principal, and the Principal said; "you know terrorists are Muslims". Why will he say that to a kid? So, I went there and confronted him, then he promised he was going to something about it.

One thing that is evident from Dovian's experience is the fact that he took bold steps to correct the stereotypes. I attribute this boldness to his immigration status in the country as further conversation revealed that Dovian has become a citizen of the United States. Yun, another immigrant who has spent just two years in Dayton, also explained his experience of name-calling at his workplace. However, the difference between Dovian and Yun was clear when I asked him what he did to correct this, and he responded that he walked away to avoid trouble. In his words: "I didn't want it to escalate or police getting involved so I chose to walk away." From the two experiences of immigrants and the attack on the mosque, it is obvious that despite the public engagement during the formation of Welcome Dayton, some native-born are still yet to come to terms with sharing their city with immigrants.

Chapter 5: Summary of Findings and Conclusion

This study has shown the role of immigrants in revitalizing Dayton, Ohio and how successful the "Welcome Dayton" initiative has been. First, the study has shown that the population and economic decline of Dayton can be attributed to the disinvestment of capital. Just like other Rust Belt cities, Dayton's decline began after the relocation of manufacturing industries to other locations leading to population loss and economic decline. Several approaches such as urban renewal, demolition, and right-sizing have been adopted to revitalize these cities. However, some, like Philadelphia's Neighborhood Transformation Initiative, have failed woefully and have even led to further out-migration. In recent times, city officials have decided to adopt smart shrinkage approaches which aim to make their current population more comfortable in the city. Becoming immigrant-friendly is one of these smart approaches and it seems to have become a trend in the redevelopment strategies of Midwest shrinking cities like Indianapolis, IN and Youngstown, OH.

In 2010, Dayton's city commissioner and director of the Human Resource Council came together after noticing significant discrimination against immigrants and discussed how the city could be more welcoming. The result of this meeting is the Welcome Dayton initiative which aimed at making immigrants feel at home. This study was able to explain the formation process, the framework of this initiative, and how it operates. Then I went further to assess how it has impacted immigrants, the city, and how residents have reacted to this new revitalization approach.

Unlike Philadelphia's NTI project which ended up not meeting its goals as a result of the government not fully involving residents in the formation stage. Dayton's initiative has shown the importance of involving residents in policy formation. In the formation stage, the director of the Human Resource Council ensured that all stakeholders—including the existing immigrant community and native residents—were involved in the conversations. This is crucial in the formation stage as it helps to identify possible resistance that could surface in the future. During this formation, no tangible resistance was recorded which suggests that residents supported this initiative. However, Shrider (2017) pointed out an important discrepancy on how Midwest cities have portrayed themselves to the media as immigrant-friendly while in the real sense, they have supported unfriendly immigrant policies at the federal level. This is evident in the 2016 presidential election when Trump, who is an advocate of unfriendly immigrant policies, was voted in by the electoral college votes from the Midwest. While the Midwest indeed voted for Trump, the votes were not distributed uniformly across all counties. For example, in 2016, Trump won in Montgomery County with a 0.7% margin. Nevertheless, this study shows that there are cases of hostility towards immigrants despite the public acceptance of the initiative during the formulation stage. This shows that some residents must have resented this initiative but would not speak up because they were only interested in the benefits that follow this initiative. This supports Shepard's (2016) argument that local immigrant-friendly policies would not exist if cities were thriving and do not need the support of immigrants.

Forming a policy isn't enough. The framework of these policies must be grounded on a path of its goal fulfillment. The Welcome Dayton Committee understood that being immigrant-friendly requires a wide spectrum of ideas from different sectors in the city. Hence, they further divided into four main sub-committees—Business and Economic Development, Local Government and Justice System, Social and Health Services, and Community, Culture, Arts, and Education—which all came up with recommendations for the city and were responsible for different immigrant-friendly programs. To reach its goal, Welcome Dayton also partnered with different voluntary organizations such as the privately-owned Legal Institutions and the Refugee Resettlement Services. These partnerships could be criticized because they put pressure on already limited resources. Likewise, the longevity of the relationship between Welcome Dayton and these sectors can be questioned because oftentimes voluntary organizations depend on external funding to reach their goals. The significant challenge of Welcome Dayton that I identified in this study is the tussle between the local immigration policy and the federal policies which frown at being immigrant friendly. The local government has been proactive in welcoming immigrants however their efforts are being undermined by federal policies. These policies have encouraged the growth of resistance groups which have further taunted the goal of Welcome Dayton.

Despite the challenges faced by Welcome Dayton, it has successfully attracted immigrants into the community and these new immigrants have started redeveloping Dayton in several ways. This study revealed that Aishka Turks have continued to increase because of the existing immigrant community they have in Dayton. This proves Portes and Rubaurt's (2006) argument that existing immigrant communities have a huge influence on future patterns of migration. Some other pull factors that were identified are the low cost of living, investment opportunities, and the presence of educational institutions. These immigrants have had both positive and surprisingly negative impacts on the city. Immigrants have contributed to the economic growth of the city through tax payments both to the federal and state government. This provides support to the Carr et. al (2012) assertion that immigrants have revived the dwindling economy of Midwest cities. Borges-Mendez (2005) identified that investment of immigrants has played a significant role in the neighborhood revitalization in East Boston. This is supported by the findings on how immigrants have helped reshaped Dayton into a better place through investing in its abandoned properties. However, their investments are concentrated in the northern part of the city leaving the western part desolated. In a real sense, immigrants cannot be blamed for this since they are at liberty to invest in any part of the city. It is only wise of them to choose where they already have an existing community. Aside from uneven development, the concentration of immigrants in the northern part has also led to a gradual change in migration pattern of immigrants within the city. Some Turkish community members are now beginning to leave Dayton for the suburbs due to the increasing cost of living in the Northern part of the city. Although their businesses are still in Dayton, this group has started forming an immigrant community which is an important factor in the context of reception outside Dayton. It is therefore important that proper planning measures are put in place to discourage this gradual suburbanization. Otherwise, the Old North Dayton might experience another shrinkage in the near future.

Some of the challenges that have hindered immigrants from integrating successfully into the city were identified in this study. A few of the struggles that surfaced from my findings are the language barrier, credit system in the United States, and expensive education. For Welcome Dayton to reach its goal, some of these challenges have been addressed through private partnerships with the existing immigrant community. An example is Dayton Public Schools which has provided educational infrastructure where new immigrants and refugees are being taught English.

In conclusion, this study has been able to provide insight on the immigrant-friendly initiatives of Dayton, how it operates, as well as offer an assessment of its benefits. This study has proved that Welcome Dayton relies on community-based organizations to reach its welcoming goal. This corroborates Shepard's (2006) finding that the success of Welcome Dayton could be attributed to the relationship and support from both the municipal government and other private organizations. Dayton's immigrant-friendly approach was said to have been adopted on the foundation of human rights. However, this study has shown that the benefits that come with being immigrant-friendly were the driving force behind the Welcome Dayton plan. The economic benefits of welcoming immigrants were used in the framework of its publicity, and acceptance across the city. This implies that, there is a symbiotic relationship between immigrants and the city where immigrants are provided a welcoming environment in exchange for rebuilding the declining economy of the city. However, Shepard (2016) mentioned that this type of relationship places the responsibility of economic revitalization on immigrants who have to succeed despite the many challenges they are faced with.

Shrider (2017) also pointed out that the success of immigrant-friendly cities has been frustrated by unfriendly policies at the federal level. This is evident in Dayton as some of these unfriendly policies have hindered the integration effort of the plan. Nevertheless, there are evident indicators that this plan has been successful in attracting immigrant populations who have helped reshaped the city economically, culturally, and environmentally. Also, this study has shown the importance of involving residents in the formation of revitalization policies most especially when the city is adopting an immigrant-friendly approach. This helps cushion the possible resistance that could emanate from the implementation of such plans. Although there is evidence that some of Dayton's residents are not as welcoming as expected, there has not been widespread public resistance to these initiatives. I therefore recommend to the Human Resource Council that they should ensure that immigrants' experiences are heard frequently in their quarterly meetings. This would not only help them evaluate the progress of the plan, but also reveal how the council can better serve both resident and immigrant communities.

Ultimately, the Welcome Dayton Plan could serve as a model for other shrinking cities that plan on adopting the immigrant-friendly strategy. However, I recommend that such cities should ensure that there is a lot of community engagement during the formation. Just like Dayton, suggestions and ideas should be welcomed from both existing immigrants and native residents because they play a key role in the context of reception of any city. Also, the findings from this study have showed the possible negative impact of this revitalization strategy. Hence it is important that cities adopting this approach should ensure that there are other revitalization strategies to cushion the

negative impacts identified in this research.

References

"Brownfield's success story" assessed from Brownfields Redevelopment Provides Improved Social Services to Immigrant Hmong Community of St. Paul, Minnesota (July 2010) (epa.gov) accessed on 3/20/2021

Adelman, Robert, Lesley Williams Reid, Gail Markle, Saskia Webb, & Charles Jaret. 2017. "Urban Crime Rates and the Changing Face of Immigration: Evidence Across Four Decades." *Journal of Ethnicity in Criminal Justice* 15(1): 52-77.

Al-Yateem, N. (2012). The effect of interview recording on quality of data obtained: A methodological reflection. *Nurse researcher*, *19*(4).

Arfanuzzaman, M., & Dahiya, B. (2019). Sustainable urbanization in Southeast Asia and beyond: Challenges of population growth, land use change, and environmental health. *Growth and Change*, *50*(2), 725-744.

Birks, M., Chapman, Y., & Francis, K. (2008). Memoing in qualitative research: Probing data and processes. *Journal of research in nursing*, *13*(1), 68-75.

Borges-Mendez, R., Liu, M., & Watanabe, P. (2005). Immigrant entrepreneurs and neighborhood revitalization: Studies of the Allston Village, East Boston and Fields Corner neighborhoods in Boston.

Bowen, Glenn A., (2009). Document Analysis as a Qualitative Research Method, *Qualitative Research Journal, Vol 9. No 2 pp. 27-40. DOI 10.3316/QRJ0902027*

Buckley, Geoffrey L. 2014. Urban Sustainability. In Cities of North America: Contemporary Challenges in U.S. and Canadian Cities. Edited by Lisa Benton-Short. Lanham: Rowman and Littlefield, pp. 377–403.

Carr, P. J., Lichter, D. T., & Kefalas, M. J. (2012). Can immigration save small-town America? Hispanic boomtowns and the uneasy path to renewal. *The Annals of the American Academy of Political and Social Science*, *641*(1), 38-57.

Cervero, R., & Landis, J. (1997). Twenty years of the Bay Area Rapid Transit system: Land use and development impacts. *Transportation Research Part A: Policy and Practice, 31*(4), 309-333.

De Graauw, Els, & Bloemaard Irene (2017): "Working Together. Building Successful Policy and Program Partnership for Immigrants" *Journal on Migration and Human Security. 5*(1), 105-123.

Duneier, M. (2016). *Ghetto: The invention of a place, the history of an idea.* Macmillan.

Emeric, E., & Newman, G., (2018). Public Transportation as Potential Remedy to Urban Decline in Dayton, Ohio. A case Study. CLEA Conference Paper No 7.

Evans, J. P. (2012). Sustainable regeneration. In H. Lovell, M. Elsenga, & S. Smith (Eds.), *International Encyclopedia of housing and home.* Elsevier.

Evans, J., & Jones, P. (2008). Rethinking sustainable urban regeneration: ambiguity, creativity, and the shared territory. *Environment and Planning A, 40*(6), 1416-1434.

Faga, B. (2006). Designing Public Consensus: the civic theater of community participation for architects, landscape architects, planners, and urban designers. Hoboken, NJ: John Wiley and Sons.

Fleury-Steiner, Benjamin, & Jamie Longazel. 2010. "Neoliberalism, Community Development, and Anti-Immigrant Backlash in Hazelton, Pennsylvania" in Monica W. Varsanyi, ed. *Taking Local Control: Immigration Policy Activism in U.S. Cities and States.* Stanford, CA: Stanford University Press.

Ghaljaie, F., Naderifar, M., & Goli, H. (2017). Snowball Sampling: A purposeful Method of Sampling in Qualitative Research. *Strides in Development of Medical Education, 14*(3), doi: 10.5812/sdme.67670

Gockowski A. (2019). "We Build the Wall, Inc. in Ohio: You, Sherred Brown, are Complicit in Death of this Lady's Daughter" The Ohio Star accessed on 01/27/2021 from We Build the Wall, Inc. in Ohio: 'You, Sherrod Brown, are Complicit in the Death of This Lady's Daughter' - The Ohio Star

Greenberg, M., Lowrie, K., Mayer, H., Miller, K. T., & Solitare, L. (2001). Brownfield redevelopment as a smart growth option in the United States. *Environmentalist, 21*(2), 129-143.

Haase, A., Rink, D., Grossmann, K., Bernt, M., & Mykhnenko, V. (2014). Conceptualizing urban shrinkage. *Environment and Planning A, 46*(7), 1519-1534.

Harvey, D. (1989). From managerialism to entrepreneurialism: the transformation in urban governance in late capitalism. *Geografiska Annaler: Series B, Human Geography, 71*(1), 3-17.

Hay, Iain (2016). *Qualitative Research Methods in Human Geography (Fourth Edition).* New York: Oxford University Press.

Heim LaFrombois, M. E., Park, Y., & Yurcaba, D. (2019). How US shrinking cities plan for change: Comparing population projections and planning strategies in depopulating US cities. *Journal of Planning Education and Research,* 0739456X19854121

Herman, Richard T., & Robert L. Smith. 2009. *Immigrant, Inc.: Why Immigrant Entrepreneurs Are Driving the New Economy (and How They Will Save the American Worker).* Hoboken, New Jersey: John Wiley & Sons, Inc.

Hollander, J. B., Pallagst, K., Schwarz, T., & Popper, F. J. (2009). Planning shrinking cities. *Progress in planning, 72*(4), 223-232.

Jones, D. (2008). *Mass Motorization & Mass Transit.* Bloomington, IN: Indiana University Press.

Joseph Schilling & Jonathan Logan (2008): "Greening the rust belt: A green Infrastructure model for right sizing America's shrinking cities" Journal of American Planning Association Vol 74 No 4 pg 451-466.

Karina Pallagst, J.V Hernandez & Patricia Hammer (2019). "Green Innovation Areas- En Route to Sustainability for Shrinking Cities?" Journal of Sustainability Vol 11, 6674 pg 1- 17.

Korkmaz, C., & Balaban, O. (2020). Sustainability of urban regeneration in Turkey: Assessing the performance of the North Ankara Urban Regeneration Project. *Habitat International, 95*, 102081.

Korkmaz, C., & Balaban, O. (2020). Sustainability of urban regeneration in Turkey: Assessing the performance of the North Ankara Urban Regeneration Project. *Habitat International, 95*, 102081.

Kunstler, J. H. (1993). *The Geography of Nowhere: The rise and decline of America's-made landscape.* New York: Simon & Schuster.

Lichfield, D. (1992). Urban Regeneration for the 1990s. *London Planning Advisory Committee, London.*

Hemphill, L., McGreal, S., & Berry, J. (2004). An indicator-based approach to measuring sustainable urban regeneration performance: Part 2, empirical evaluation and case-study analysis. *Urban Studies, 41*(4), 757-772.

Long, L., & Nucci, A. (1997). The 'clean break' revisited: is US population again deconcentrating?. *Environment and Planning A, 29*(8), 1355-1366.

Maliene, V., Wignall, L., & Malys, N. (2012). Brownfield regeneration: Waterfront site developments in Liverpool and Cologne. *Journal of Environmental Engineering and Landscape Management, 20*(1), 5-16.

Martinez-Fernandez, C., Audirac, I., Fol, S., & Cunningham-Sabot, E. (2012). Shrinking cities: Urban challenges of globalization. *International journal of urban and regional research, 36*(2), 213-225.

Martinez-Fernandez, C., Hinojosa, C., & Miranda, G. (2010). Greening Jobs and Skills. Development (LEED): Paris, France, 2012; pp. 41–46.

Martinez-Fernandez, C., Weyman, T., Fol, S., Audirac, I., Cunningham-Sabot, E., Wiechmann, T., & Yahagi, H. (2016). Shrinking cities in Australia, Japan, Europe and the USA: From a global process to local policy responses. *Progress in Planning, 105*, 1-48.

McGovern, S. J. (2006). Philadelphia's neighborhood transformation initiative: A case study of mayoral leadership, bold planning, and conflict. *Housing Policy Debate*, *17*(3), 529-570.

Metzger, J. T. (2000). Planned abandonment: The neighborhood life-cycle theory and national urban policy. *Housing Policy Debate* Vol 11(1) pp 7-40.

Millsap Adam (2019); Polishing the Gem City: Dayton, Ohio's Rise, Decline, and Transition. https://www.forbes.com/sites/adammillsap/2019/11/11/polishing-the-gem-city-dayton-ohios-rise-decline-and-transition/#56297842577f accessed 19th Nov. 2019

Miranda C Hallet and Theo J. Majika (2019): "The Benefits that Places like Dayon Ohio Reap by Welcoming Immigrants". http://theconversation.com/the-benefits-that-places-like-dayton-ohio-reap-by-welcoming-immigrants-107949 accessed 20th Nov. 2019

New American Economy Report (2020) accessed from New Americans in Dayton, Ohio - New American Economy Research Fund accessed on the 1st Mar. 2021

Oswalt, P., & Rieniets, T. (2006). *Atlas of shrinking cities*.

Martinez- Fernandez, C., Kubo, N., Nova, A., Pallagst, K., Fleschurz, R., & Said, S. (2017). What drives planning in a shrinking city? Tales from two German and two American cases. *Town Planning Review*, *88*(1), 15-28.

Pallagst, K., Fleschurz, R., & Trapp, F. (2017). Greening the shrinking city—policies and planning approaches in the USA with the example of Flint, Michigan. *Landscape research*, *42*(7), 716-727.

Pallagst, K., Wiechmann, T., & Martinez-Fernandez, C. (Eds.). (2013). *Shrinking cities: international perspectives and policy implications*. Routledge.

Plevoets, B., & Sowińska-Heim, J. (2018). Community initiatives as a catalyst for regeneration of heritage sites: Vernacular transformation and its influence on the formal adaptive reuse practice. *Cities*, *78*, 128-139.

Portes, Alejandro, & Ruben G. Rumbaut. 2006. *Immigrant America: A Portrait.* Third Edition. Berkeley, CA: University of California Press.

Preston Julia (2013): "Ailing Cities Across the Midwest Extend A Welcoming Hand to Immigrants" A New York Time's Post Published Oct 7, 2013. Pg A18.

Reimer, M., Rusche, K., & Othengrafen, F. (2015). Grüne Infrastruktur in urbanen Räumen. *RaumPlanung, 4,* 6-7.

Rhodes, J., & Russo, J. (2013). Shrinking 'smart'?: Urban redevelopment and shrinkage in Youngstown, Ohio. *Urban Geography, 34*(3), 305-326.

Rink, D., Haase, A., Grossmann, K., Couch, C., & Cocks, M. (2012). From long-term shrinkage to re-growth? The urban development trajectories of Liverpool and Leipzig. *Built environment, 38*(2), 162-178.

Roberts, I. (2000). Leicester environment city: learning how to make Local Agenda 21, partnerships and participation deliver. *Environment and urbanization, 12*(2), 9-26.

Rogerson, P. A., & Plane, D. A. (2013). The Hoover index of population concentration and the demographic components of change: An article in memory of Andy Isserman. *International Regional Science Review, 36*(1), 97-114.

Rybczynski, W., & Linneman, P. D. (1999). How to save our shrinking cities. *Public Interest,* 30-44.

Haase, A., Rink, D., Grossmann, K., Bernt, M., & Mykhnenko, V. (2014). Conceptualizing urban shrinkage. *Environment and Planning A, 46*(7), 1519-1534.

Rybczynski, Witold, & Peter D. Linneman. 1999. "How to Save Our Shrinking Cities." *Public Interest* 135:30–44.

Sahatcija, R., Ferhataj, A., & Ora, A. (2020). Push-Pull Factors of Migration Today in Albania. *Journal of International Cooperation and Development, 3*(1), https://doi.org/10.36941/jicd-2020-0001

Sanders Thompson, V., Wells, A., Eds.; Lexington Books: Lanham, MD, USA, 2013; pp. 241–268. ISBN 978-0739177006.

Schilling, J., & Logan, J. (2008). Greening the rust belt: A green infrastructure model for right-sizing America's shrinking cities. *Journal of the American Planning Association, 74*(4), 451-466. EPA 2010.

Shepard, A., (2016). The Emerging "Immigrant-Friendly" City: How and why cities frame themselves as welcoming places to immigrants. (Unpublished MSc book) University of Vermont.

Shrider, E., (2017). Can Immigrants Save the Rust Belt? Struggling Cities, Immigration, and Revitalization (Unpublished doctoral book) Ohio State University.

Singer, Audrey, and Jill H. Wilson. 2006. "From 'There' to 'Here': Refugee Resettlement in Metropolitan America." *The Living Cities Census Series*. Washington, DC: Brookings Institution.

Smith, N. (1982). Gentrification and uneven development. *Economic geography, 58*(2), 139-155.

Sugrue, T. (1996). *The Origins of the Urban Crisis: Race and Inequality in Postwar Detroit*. Princeton, NJ: Princeton University Press.

Tallon, A. (2010). Urban renaissance in the UK: city responses and critical issues. In *Policy, Legislation and Implementation in Urban Regeneration*.

Vigdor, J. L., Calcagni, G., & Valvekhar, M. (2013). Immigration and the Revival of American Cities: From Preserving Manufacturing Jobs to Strengthening the Housing Market. Washington, DC: Partnership for a New American Economy.

Welcome Dayton. (2016). Immigrants of Dayton. Retrieved from Immigrants of Dayton - Welcome Dayton - Immigrant Friendly City | Welcome Dayton – Immigrant Friendly City accessed on 12th Dec. 2020

Wheeler, B. K., & Beatley, T. (2008). © 2008. *The sustainable urban development reader*.

Wiechmann, T., & Pallagst, K. M., 2012, "Urban shrinkage in Germany and the USA: a comparison of transformation patterns and local strategies" *International Journal of Urban and Regional Research* 36: 261–280

Wiechmann, T. (2008). Errors expected—aligning urban strategy with demographic uncertainty in shrinking cities. *International Planning Studies*, *13*(4), 431-446.

Wiechmann, T., & Pallagst, K. M. (2012). Urban shrinkage in Germany and the USA: A comparison of transformation patterns and local strategies. *International journal of urban and regional research*, *36*(2), 261-280.

Appendix A: Recruitment Document

I am Adeuga, Adewole Michael. An MSc student of Geography at Ohio University. I am currently working on my book which is focused on the revitalization of Dayton. I recently had an interview session with .... who recommended that I speak with you. Please see a brief overview of the research below.

In the United States, numerous cities have experienced decline; however, some have developed policies to regenerate the dying central city. Many of these cities are actively combating the decline by launching creative initiatives aimed at addressing vacant and abandoned properties. This study aims to assess the sustainability goals of municipal immigration policies using Dayton, Ohio as a case study. Also, this research aims to assess the impact recent immigrants have had on Dayton and the attitude of Daytons residents towards these immigrants. This project intends to provide answers to how city officials have been able to incorporate revitalization goals into the city's sustainable goal and also reveal if truly these revitalization plans have been successful so far. Hence conclude if truly immigration could be used as a redevelopment tool by other shrinking cities.

Participants are expected to have at most an hour interview with researchers were questions regarding welcoming Dayton and how immigrants have helped reshaped the city would be asked. All interviews will be semi-structured to enable the interview questions to be informed by the perspective of the participants. Given the current pandemic situation, interviews would only be conducted online via skype teems or any other virtual supporting software at participants' convenience. Interviews will be recorded with a digital recorder and will subsequently be transcribed. It should be noted that no identifying information will be asked or recorded.

If you have any questions regarding this study, kindly contact the investigator *Adeuga, Adewole Michael at aa075619@ohio.edu, or (301)458-9703* or the advisor *Harold Perkins, at perkinsh@ohio.edu or (740) 593-9896*.

Appendix B: Interview Protocol

Decision Makers

Introduction

- Current/Past position

- Level of involvement in the formation of policies

What exactly gave the need for the establishment of Welcome Dayton

- Why did Dayton decide to take this (immigration) route in the revitalization

  process?

What were the processes involved in the formation of policies to redevelop the city?

- How were the committee members selected? Were all groups represented during

  this formation?

- Did it include immigrants/Planners and other key decision-makers?

How do these initiatives operate?

- What are the processes involved in integrating immigrants into the city?

What is the resistance this initiative has encountered?

- How did the native residents react to this initiative in 2011 when it was established?

- Given the current immigration policy at the federal government level, how have

  city officials tried to balance the goal of these initiatives?

Why do you think a lot of immigrants are migrating to Dayton instead of other cities in the

US?

- How affordable is Dayton compared to other neighboring cities?

How have these immigrants helped reshaped the community?

- Economically and Environmentally

Do you think Welcome Dayton has impacted immigrants?

How does Welcome Dayton keep track of new immigrants within the city?

Is there any aspect of Welcome Dayton that could be adjusted?

Closing interview.

- Would you like to say more about the redevelopment policies of Dayton?

- Do you have any questions for me?

- Thanks for participating.

Interview Protocol-immigrants

Introduction

Tell me about yourself

- Family

- Level of Education- Did you study in the United States?

- Occupation- How long have you been on the job?

- How long have you been living in Dayton?

What were the pull factors that influenced your relocation to Dayton?

- Did Welcome Dayton influence your decision to move to Dayton?

- Have you heard of any Dayton initiative which supports immigrants?

Could you please describe what Dayton looked like as at the time you moved in?

How has it faired since you moved in have you noticed any development?

- What are the significant changes you have seen in housing development?

Can you compare your past environment/city to Dayton?

- Cost of living in Dayton compared to previous city.

How have you been able to integrate within the community given the different cultural backgrounds?

- Are you a member of any community group which shares the same cultural values as you?
    - o How has this group been beneficial to you?
    - o How does this group help integrate immigrants?

What is your relationship with native residents?

- How often do you interact with them?
- Were they beneficial to you during your integration process?

What are some of the things you like about Dayton that are not in your previous city?

How have you contributed to the economic growth of the city?

Would you encourage people to move into Dayton or other suburb counties?

Closing Interview

- Is there anything else you would like to say about your experiences?
- Do you have any questions for me?

Appendix C: Public Invitation to the Welcome Dayton Formation Forum

**Dayton Human Relations Council Invites Community Dialogue
On How to be an Immigrant Friendly City**

Release Date: Wednesday, March 30, 2011

Contact: Tom Wahlrab, Executive Director of the Human Relations Council, 333-1403

The recent Census results indicate that Dayton's population makeup continues to evolve. To capitalize on this trend as a way to enhance the vibrancy and growth of the community, Dayton's Human Relations Council is leading efforts to help make Dayton a more immigrant friendly city.

The public is invited to join the conversation about making Dayton more friendly to immigrant groups by attending a workshop on April 5 from 9:30-11:30 a.m. at the Sinclair Conference Center. Interested attendees can register at www.mvfairhousing.com.

"Creating an immigrant-friendly community can help attract and retain citizens who are engaged in the community and eager to support neighborhood revitalization and economic development efforts," said Tom Wahlrab, Executive Director of the Human Relations Council. "Harnessing the energy, skills and resources of these new citizen groups can be an important element of a community's success."

Wahlrab said the Immigrant Friendly City initiative is designed to create a conversation with community members who are actively engaged in supporting the integration of new residents. The purpose is to:

- propose City policies to support integration of new residents and further business development;

- explore support systems that engage new residents in civic affairs and lead to full citizenship;

- gain collaborative commitments to work together; and

- envision the future of our city as an immigrant friendly city.

Unlike in the past, when most immigrant groups were concentrated in a few states, significant concentrations of immigrants now exist in cities across the country.

Wahlrab highlighted several key benefits of being known as an immigrant friendly community:

1. Immigrant communities occupy and develop the housing stock.

2. Immigrant communities build self-governing communities.

3. Immigrant communities create businesses to serve their own and the greater community's needs, enriching the economic climate of the city.

4. Immigrant connections can open doors to new economic markets.

5. Population growth increases consumption and fuels the creation of more jobs.

6. Immigrants welcomed and accepted into the community are more willing to create and maintain safer communities.

7. Cultural diversity is broadened, thus energizing the greater community through openness and cross-cultural dialogue.

Anyone who would like to be involved in this conversation should contact Brenda Hawes with the Human Relations Council at 333-1400, or by email at Brenda.hawes@daytonohio.gov.

9 783384 259868